Numerical Analysis

Numerical Analysis

Stella Lee

CLANRYE
INTERNATIONAL
www.clanryeinternational.com

Clanrye International,
750 Third Avenue, 9th Floor,
New York, NY 10017, USA

ISBN: 978-1-64726-101-6

Cataloging-in-Publication Data

Numerical analysis / Stella Lee.
p. cm.
Includes bibliographical references and index.
ISBN: 978-1-64726-101-6
1. Numerical analysis. 2. Mathematical analysis. I. Lee, Stella.
QA297 .N86 2022
518--dc23

For information on all Clanrye International publications visit our website at www.clanryeinternational.com

TABLE OF CONTENTS

PREFACE

The purpose of this book is to help students understand the fundamental concepts of this discipline. It is designed to motivate students to learn and prosper. I am grateful for the support of my colleagues. I would also like to acknowledge the encouragement of my family.

The study of algorithms that use numerical approximation for the problems of mathematical analysis is known as numerical analysis. Some of the major areas of study within this field are computing values of functions, interpolation, extrapolation, regression, optimization, evaluating integrals and differential equations. The major methods that are used in numerical analysis are direct and iterative methods. Direct methods compute the solution to a problem in a finite number of steps and are able to give the answer if they are performed in infinite precision arithmetic. Iterative methods form successive approximations that converge to the exact solution only in the limit. Numerical analysis is applied in all fields of physical sciences and engineering as well as in social sciences, medicine and life sciences. This book unravels the recent studies in the field of numerical analysis. The topics included herein on numerical analysis are of utmost significance and bound to provide incredible insights to readers. Those in search of information to further their knowledge will be greatly assisted by this book.

A foreword for all the chapters is provided below:

Chapter - Numerical Analysis: An Introduction

Numerical analysis is the branch of mathematics that deals with the use of numerical approximations for solutions to mathematical problems and analysis. Numerical method and iterative method are two types of methods that are used in numerical analysis. This chapter sheds light on numerical analysis for a thorough understanding of the subject.

Chapter - Numerical Differentiation and Integration

Numerical differentiation refers to the algorithms that are used for finding the derivative of a mathematical function using values of a function at a given point. Numerical integration uses a wide range of algorithms to estimate the numerical value of an integral and describe the numerical solution of differential equations. This chapter discusses numerical differentiation and integration in detail.

Chapter - Algorithms in Numerical Analysis

There are various algorithms that are used in numerical analysis such as Brunn's FFT algorithm, Goertzel algorithm, prime factor FFT algorithm, vector-radix FFT algorithm, arbitrary-precision arithmetic, Durand-Kerner method, Brent's method, etc. The topics elaborated in this chapter will help in gaining a better perspective of the different algorithms used in numerical analysis.

Chapter - Numerical Linear Algebra

Numerical linear algebra is the study of matrix operations for creating computer algorithms that provide efficient solutions to mathematical problems. Bartels-Stewart algorithm, GotoBLAS, DADiSP, LAPACK, MATLAB, Armadillo, etc. are studied under its domain. This chapter has been carefully written to provide an easy understanding of numerical linear algebra.

Chapter - Mathematical Optimization

Mathematical optimization refers to a collection of mathematical principles and methods to solve optimization problems. Linear and non-linear programming, and dynamic programming are also studied under it. This chapter closely examines the aspects related to mathematical optimization for an in-depth understanding of the subject.

Stella Lee

Numerical Analysis: An Introduction

CHAPTER 1

Numerical analysis is the branch of mathematics that deals with the use of numerical approximations for solutions to mathematical problems and analysis. Numerical method and iterative method are two types of methods that are used in numerical analysis. This chapter sheds light on numerical analysis for a thorough understanding of the subject.

Numerical analysis is an area of mathematics and computer science that creates, analyses and implements algorithms for obtaining numerical solutions to problems involving continuous variables. Such problems arise throughout the natural sciences, social sciences, engineering, medicine, and business.

Modern Applications and Computer Software

Numerical analysis and mathematical modeling are essential in many areas of modern life. Sophisticated numerical analysis software is commonly embedded in popular software packages (e.g., spreadsheet programs) and allows fairly detailed models to be evaluated, even when the user is unaware of the underlying mathematics. Attaining this level of user transparency requires reliable, efficient, and accurate numerical analysis software, and it requires problem-solving environments (PSE) in which it is relatively easy to model a given situation. PSEs are usually based on excellent theoretical mathematical models, made available to the user through a convenient graphical user interface.

Applications

Computer-aided engineering (CAE) is an important subject within engineering, and some quite sophisticated PSEs have been developed for this field. A wide variety of numerical analysis techniques is involved in solving such mathematical models. The models follow the basic Newtonian laws of mechanics, but there is a variety of possible specific models, and research continues on their design. One important CAE topic is that of modeling the dynamics of moving mechanical systems, a technique that involves both ordinary differential equations and algebraic equations (generally nonlinear). The numerical analysis of these mixed systems, called differential-algebraic systems, is quite difficult but necessary in order to model moving mechanical systems. Building simulators for cars, planes, and other vehicles requires solving differential-algebraic systems in real time.

Another important application is atmospheric modeling. In addition to improving weather forecasts, such models are crucial for understanding the possible effects of human activities on the Earth's climate. In order to create a useful model, many variables must be introduced. Fundamental among these are the velocity $V(x, y, z, t)$, pressure $P(x, y, z, t)$, and temperature $T(x, y, z, t)$, all given at position (x, y, z) and time t. In addition, various chemicals exist in the atmosphere, including ozone, certain chemical pollutants, carbon dioxide, and other gases and particulates, and their interactions have to be considered. The underlying equations for studying $V(x, y, z, t)$, $P(x, y, z, t)$, and $T(x, y, z, t)$ are partial differential equations; and the interactions of the various chemicals are described using some quite difficult ordinary differential equations. Many types of numerical analysis procedures are used in atmospheric modeling, including computational fluid mechanics and the numerical solution of differential equations. Researchers strive to include ever finer detail in atmospheric models, primarily by incorporating data over smaller and smaller local regions in the atmosphere and implementing their models on highly parallel supercomputers.

Modern businesses rely on optimization methods to decide how to allocate resources most efficiently. For example, optimization methods are used for inventory control, scheduling, determining the best location for manufacturing and storage facilities, and investment strategies.

Computer Software

Software to implement common numerical analysis procedures must be reliable, accurate, and efficient. Moreover, it must be written so as to be easily portable between different computer systems. Since about 1970, a number of government-sponsored research efforts have produced specialized, high-quality numerical analysis software.

The most popular programming language for implementing numerical analysis methods is Fortran, a language developed in the 1950s that continues to be updated to meet changing needs. Other languages, such as C, C++, and Java, are also used for numerical analysis. Another approach for basic problems involves creating higher level PSEs, which often contain quite sophisticated numerical analysis, programming, and graphical tools. Best known of these PSEs is MATLAB, a commercial package that is arguably the most popular way to do numerical computing. Two popular computer programs for handling algebraic-analytic mathematics (manipulating and displaying formulas) are Maple and Mathematica.

GENERATION AND PROPAGATION OF ERRORS

Propagation of errors is essential to understanding how the uncertainty in a parameter affects computations that use that parameter. The uncertainty propagates by a set of rules into your solution. These rules are not easy to remember, or apply to complicated situations, and are only approximate for equations that are nonlinear in the parameters.

We will use a Monte Carlo simulation to illustrate error propagation. The idea is to generate a distribution of possible parameter values, and to evaluate your equation for each parameter value. Then, we perform statistical analysis on the results to determine the standard error of the results.

We will assume all parameters are defined by a normal distribution with known mean and standard deviation.

Addition and Subtraction

```
import numpy as np

import matplotlib.pyplot as plt


N = 1e6 # number of samples of parameters


A_mu = 2.5; A_sigma = 0.4

B_mu = 4.1; B_sigma = 0.3


A = np.random.normal(A_mu, A_sigma, size=(1, N))

B = np.random.normal(B_mu, B_sigma, size=(1, N))


p = A + B

m = A - B


print np.std(p)

print np.std(m)


print np.sqrt(A_sigma**2 + B_sigma**2) # the analytical std dev

>>> >>> >>> >>> >>> >>> >>> >>> >>> >>> >>> >>> >>> 0.500505424616

0.500113385681
```

```
>>> 0.5
```

Multiplication

```
F_mu = 25.0; F_sigma = 1;

x_mu = 6.4; x_sigma = 0.4;


F = np.random.normal(F_mu, F_sigma, size=(1, N))

x = np.random.normal(x_mu, x_sigma, size=(1, N))


t = F * x

print np.std(t)

print np.sqrt((F_sigma / F_mu)**2 + (x_sigma / x_mu)**2) * F_mu * x_mu

>>> >>> >>> >>> >>> >>> 11.8900166284

11.8726576637
```

Division

This is really like multiplication: F / x = F * (1 / x).

```
d = F / x

print np.std(d)

print np.sqrt((F_sigma / F_mu)**2 + (x_sigma / x_mu)**2) * F_mu / x_mu

0.293757533168

0.289859806243
```

Exponents

This rule is different than multiplication (A^2 = A*A) because in the previous examples we assumed the errors in A and B for A*B were uncorrelated. In A*A, the errors are not uncorrelated, so there is a different rule for error propagation.

```
t_mu = 2.03; t_sigma = 0.01*t_mu; # 1% error
```

```
A_mu = 16.07; A_sigma = 0.06;

t = np.random.normal(t_mu, t_sigma, size=(1, N))
A = np.random.normal(A_mu, A_sigma, size=(1, N))

# Compute t^5 and sqrt(A) with error propagation
print np.std(t**5)
print (5 * t_sigma / t_mu) * t_mu**5
>>> >>> >>> >>> >>> ... 1.72454836176
1.72365440621
print np.std(np.sqrt(A))
print 1.0 / 2.0 * A_sigma / A_mu * np.sqrt(A_mu)
0.00748903477329
0.00748364738749
```

Chain Rule in Error Propagation

Let v = vo + a*t, with uncertainties in vo, a and t:

```
vo_mu = 1.2; vo_sigma = 0.02;
a_mu = 3.0;  a_sigma  = 0.3;
t_mu = 12.0; t_sigma  = 0.12;

vo = np.random.normal(vo_mu, vo_sigma, (1, N))
a = np.random.normal(a_mu, a_sigma, (1, N))
t = np.random.normal(t_mu, t_sigma, (1, N))

v = vo + a*t

print np.std(v)
```

```
print np.sqrt(vo_sigma**2 + t_mu**2 * a_sigma**2 + a_mu**2 * t_sig-
ma**2)

>>> >>> >>> >>> >>> >>> >>> >>> >>> 3.62232509326

3.61801050303
```

AFFINE ARITHMETIC

Affine arithmetic is a model for self-validated numerical analysis. In AA, the quantities of interest are represented as affine combinations of certain primitive variables, which stand for sources of uncertainty in the data or approximations made during the computation. Affine arithmetic is meant to be an improvement on interval arithmetic, and is similar to generalized interval arithmetic, first-order Taylor arithmetic, the center-slope model, and ellipsoid calculus – in the sense that it is an automatic method to derive first-order guaranteed approximations to general formulas. Affine arithmetic is potentially useful in every numeric problem where one needs guaranteed enclosures to smooth functions, such as solving systems of non-linear equations, analyzing dynamical systems, integrating functions differential equations, etc. Applications include ray tracing, plotting curves, intersecting implicit and parametric surfaces, error analysis, process control, worst-case analysis of electric circuits, and more.

NUMERICAL METHOD

Numerical methods designed for the constructive solution of mathematical problems requiring particular numerical results, usually on a computer. A numerical method is a complete and unambiguous set of procedures for the solution of a problem, together with computable error estimates.

Development of Numerical Methods

Numerical analysts and applied mathematicians have a variety of tools which they use in developing numerical methods for solving mathematical problems. An important perspective, which cuts across all types of mathematical problems is that of replacing the given problem with a 'nearby problem' which can be solved more easily. There are other perspectives which vary with the type of mathematical problem being solved.

Numerical Solution of Systems of Linear Equations

Linear systems arise in many of the problems of numerical analysis, a reflection of the approximation of mathematical problems using linearization. This leads to diversity in the characteristics of linear systems, and for this reason there are numerous approaches to solving linear systems. As an example, numerical methods for solving partial differential equations often lead to very large 'sparse' linear systems in which most coefficients are zero. Solving such sparse systems requires methods that are quite different from those used to solve more moderate sized 'dense' linear systems in which most coefficients are non-zero.

There are 'direct methods' and 'iterative methods' for solving all types of linear systems, and the method of choice depends on the characteristics of both the linear system and on the computer hardware being used. For example, some sparse systems can be solved by direct methods, whereas others are better solved using iteration. With iteration methods, the linear system is sometimes transformed to an equivalent form that is more amenable to being solved by iteration; this is often called 'pre-conditioning' of the linear system.

With the matrix eigenvalue problem $Ax = \lambda x$, it is standard to transform the matrix A to a simpler form, one for which the eigenvalue problem can be solved more easily and cheaply. A favorite choice are 'orthogonal transformations' because they are a simple and stable way to convert the given matrix A. Orthogonal transformations are also very useful in transforming other problems in numerical linear algebra. Of particular importance in this regard is the least squares solution of over-determined linear systems.

The linear programming problem was solved principally by the 'simplex method' until new approaches were developed in the 1980s, and it remains an important method of solution. The simplex method is a direct method that uses tools from the numerical solution of linear systems.

Numerical Solution of Systems of Nonlinear Equations

With a single equation $f(x)=0$, and having an initial estimate x_0 of the root α, approximate $f(x)$ by its tangent line at the point $(x_0, f(x_0))$. Find the root of this tangent line as an approximation to the root α of the original equation $f(x) = 0$. This leads to 'Newton's iteration method',

$$x_{n+1} = x_n - \frac{f(x_n)}{f'(x_n)}, \; n = 0,1,\ldots$$

Other linear and higher degree approximations can be used, and these lead to alternative iteration methods. An important derivative-free approximation of Newton's method is the 'secant method'.

For a system of m nonlinear equations for a solution vector x in R^m, we approximate $f(x)$

by its linear Taylor approximation about the initial estimate x_0. This leads to Newton's method for nonlinear systems,

$$x_{n+1} = x_n - \left[f'(x_n)\right]^{-1} f(x_n),\ n = 0,1,\ldots$$

in which $f'(x)$ denotes the Jacobian matrix, of order $m \times m$ for $f(x)$.

In practice, the Jacobian matrix for $f(x)$ is often too complicated to compute directly; instead the partial derivatives in the Jacobian matrix are approximated using 'finite differences'. This leads to a 'finite difference Newton method'. As an alternative strategy and in analogy with the development of the secant method for the single variable problem, there is a similar rootfinding iteration method for solving nonlinear systems. It is called 'Broyden's method' and it uses finite difference approximations of the derivatives in the Jacobian matrix, avoiding the evaluation of the partial derivatives of $f(x)$.

Numerical Methods for Solving Differential and Integral Equations

With such equations, there are usually at least two general steps involved in obtaining a nearby problem from which a numerical approximation can be computed; this is often referred to as 'discretization' of the original problem. The given equation will have a domain on which the unknown function is defined, perhaps an interval in one dimension and maybe a rectangle, ellipse, or other simply connected bounded region in two dimensions. Many numerical methods begin by introducing a mesh or grid on this domain, and the solution is to be approximated using this grid. Following this, there are several common approaches.

One approach approximates the equation with a simpler equation defined on the mesh. For example, consider approximating the boundary value problem:

$$\begin{aligned} &u''(s) = f(s, u(s)),\ 0 \le s \le 1 \\ &u(0) = u(1) = 0. \end{aligned}$$

Introduce a set of mesh points $s_j = jh, = 0,1,\ldots,n$, with $h = 1/n$ for some given $n \ge 2$. Approximate the boundary value problem by:

$$\begin{aligned} &\frac{1}{h^2}[\tilde{u}_n(s_{j+1}) - 2\tilde{u}_n(s_j) + \tilde{u}_n(s_{j-1})] = f(s_j, \tilde{u}_n(s_j)),\ j = 1,\ldots,n-1 \\ &\tilde{u}_n(0) = \tilde{u}_n(1) = 0 \end{aligned}$$

The second derivative in the original problem has been replaced by a numerical approximation to the second derivative. The new problem is a finite system of nonlinear equations, presumably amenable to solution by known techniques. The solution to this new problem is $\tilde{u}_n$, and it is defined on only the mesh points $\{s_j\}$.

A second approach to discretizing differential and integral equations is as follows. Choose a finite-dimensional family of functions, denoted here by F, with which to approximate the unknown solution function u. Write the given differential or integral equation as $L(u)=0$, with $L(v)$ a function for any function v, perhaps over a restricted class of functions v. The numerical method consists of selecting a function $\tilde{u} \in F$ such that $L(\tilde{u})$ is a small function in some sense. The various ways of doing this lead to 'Galerkin methods', 'collocation methods', and 'least square methods'.

Yet another approach is to reformulate the equation $L(u)=0$ as an optimization problem. Such reformulations are a part of the classical area of mathematics known as the 'calculus of variations', a subject that reflects the importance in physics of minimization principles. The well-known 'finite element method' for solving elliptic partial differential equations is obtained in this way, although it often coincides with a Galerkin method.

The approximating functions in F are often chosen as piecewise polynomial functions which are polynomial over the elements of the mesh chosen earlier. Such methods are sometimes called 'local methods'. When the approximating functions $p \in F$ are defined without reference to a grid, then the methods are sometimes called 'global methods' or 'spectral methods'. Examples of such F are sets of polynomials or trigonometric functions of some finite degree or less.

With all three approaches to solving a differential or integral equations, the intent is that the resulting solution $\tilde{u}$ be close to the desired solution u. The business of theoretical numerical analysis is to analyze such an algorithm and investigate the size of $u-\tilde{u}$.

ITERATIVE METHOD

The Iterative Method is a mathematical way of solving a problem which generates a sequence of approximations. This method is applicable for both linear and nonlinear problems with large number of variables. The word Iterative or Iteration refers to the technique that solve any linear system problems with successive approximation at each step. There are two methods under iterative methods one is stationary iterative method and another is a non stationary Iterative method.

Let the given equation be f(x) = 0 and the value of x to be determined. By using the Iteration method you can find the roots of the equation. To find the root of the equation first we have to write equation like below:

$$x = pi(x)$$

Let $x = x_0$ be an initial approximation of the required root α then the first approximation x_1 is given by $x_1 = pi(xo)$.

Similarly for second, thrid and so on. Approximation:

$$x_2 = pi(x_1)$$

$x_3 = pi(x_2)$

$x_4 = pi(x_3)$

$x_n = pi(x_{n-1})$

Iteration Method

Example: Find the real root of the equation $x^3 + x^2 = 1$ by iteration method.

Solution:

We can rewrite the above equation by:

$x^3 + x^2 - 1 = 0;$

Let $f(x) = x^3 + x^2 - 1$

$f(0) = -1$ (positive)

$f(1) = 1$ (negative)

Hence the root value lie between 0 to 1.

$x^3 + x^2 - 1 = 0$

$x^2 (x + 1) = 1$

$x^2 = 1/ (x + 1)$

$$x = 1/\sqrt{(x+1)}$$

$$pi(x) = 1/\sqrt{(x+1)}$$

Let the initial approximation be $x_0 = 0.5$:

$$x_1 = pi(x_0) = 1/\sqrt{1+0.5} = 0.81649$$

$$x_2 = pi(x_1) = 1/\sqrt{1+0.81649} = 0.74196$$

$$x_3 = pi(x_2) = 1/\sqrt{1+0.74196} = 0.75767$$

$$x_4 = pi(x_3) = 1/\sqrt{1+0.75767} = 0.75427$$

$$x_5 = pi(x_4) = 1/\sqrt{1+0.75427} = 0.75500$$

$$x_6 = pi(x_5) = 1/\sqrt{1+0.75500} = 0.75485$$

$$x_7 = pi(x_6) = 1/\sqrt{1+0.75485} = 0.75488$$

Since the difference between x_6 and x_7 are very small, so the root is 0.75488.

References

- Numerical-analysis, science: britannica.com, Retrieved 14 May, 2019
- Numerical-propagation-of-errors: kitchingroup.cheme.cmu.edu, Retrieved 17 July, 2019
- Affine-arithmetic: definitions.net, Retrieved 07 June, 2019
- Numerical-analysis: scholarpedia.org, Retrieved 25 February, 2019
- Iteration-method-numerical-analysis, math-worksheets: ncalculators.com, Retrieved 28 July, 2019

Numerical Differentiation and Integration

CHAPTER 2

Numerical differentiation refers to the algorithms that are used for finding the derivative of a mathematical function using values of a function at a given point. Numerical integration uses a wide range of algorithms to estimate the numerical value of an integral and describe the numerical solution of differential equations. This chapter discusses numerical differentiation and integration in detail.

NUMERICAL DIFFERENTIATION

Numerical differentiation is the process of finding the numerical value of a derivative of a given function at a given point. In general, numerical differentiation is more difficult than numerical integration. This is because while numerical integration requires only good continuity properties of the function being integrated, numerical differentiation requires more complicated properties such as Lipschitz classes. Numerical differentiation is implemented as ND[*f*, *x*, *x0*, Scale -> *scale*] in the Wolfram Language package Numerical Calculus.

There are many applications where derivatives need to be computed numerically. The simplest approach simply uses the definition of the derivative:

$$f'(x) = \lim_{h \to 0} \frac{f(x+h) - f(x)}{h},$$

for some small numerical value of $h \ll 1$.

Finite Differences

The simplest method is to use finite difference approximations.

A simple two-point estimation is to compute the slope of a nearby secant line through the points $(x, f(x))$ and $(x + h, f(x + h))$. Choosing a small number h, h represents a small change in x, and it can be either positive or negative. The slope of this line is:

$$\frac{f(x+h) - f(x)}{h}.$$

This expression is Newton's difference quotient (also known as a first-order divided difference).

The slope of this secant line differs from the slope of the tangent line by an amount that is approximately proportional to h. As h approaches zero, the slope of the secant line approaches the slope of the tangent line. Therefore, the true derivative of f at x is the limit of the value of the difference quotient as the secant lines get closer and closer to being a tangent line:

$$f'(x) = \lim_{h \to 0} \frac{f(x+h) - f(x)}{h}.$$

Since immediately substituting 0 for h results in division by zero, calculating the derivative directly can be unintuitive.

Equivalently, the slope could be estimated by employing positions $(x - h)$ and x.

Another two-point formula is to compute the slope of a nearby secant line through the points $(x - h, f(x - h))$ and $(x + h, f(x + h))$. The slope of this line is:

$$\frac{f(x+h) - f(x-h)}{2h}.$$

This formula is known as the symmetric difference quotient. In this case the first-order errors cancel, so the slope of these secant lines differ from the slope of the tangent line by an amount that is approximately proportional to h^2. Hence for small values of h this is a more accurate approximation to the tangent line than the one-sided estimation. However, although the slope is being computed at x, the value of the function at x is not involved.

The estimation error is given by:

$$R = \frac{-f^{(3)}(c)}{6} h^2,$$

where c is some point between $x - h$ and $x + h$. This error does not include the rounding error due to numbers being represented and calculations being performed in limited precision.

The symmetric difference quotient is employed as the method of approximating the derivative in a number of calculators, including TI-82, TI-83, TI-84, TI-85, all of which use this method with $h = 0.001$.

Floating-point Finite Differences

An important consideration in practice when the function is calculated using floating-point arithmetic is how small a value of step h to choose. If chosen too small, the subtraction will yield a large rounding error. In fact, all the finite-difference formulae are ill-conditioned and due to cancellation will produce a value of zero if h is small enough. If too large, the calculation of the slope of the secant line will be more accurately calculated, but the estimate of the slope of the tangent by using the secant could be worse.

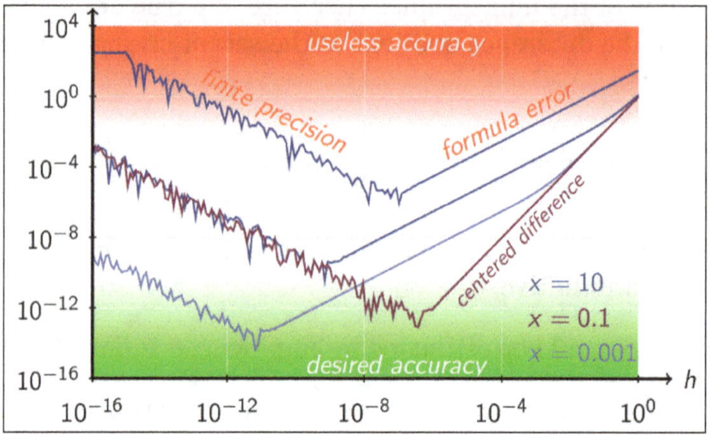

Example showing the difficulty of choosing $\hbar$ due to both rounding error and formula error.

For the numerical derivative formula evaluated at x and $x + h$, a choice for h that is small without producing a large rounding error is $\sqrt{\varepsilon}x$ (though not when $x = 0$), where the machine epsilon ε is typically of the order of 2.2×10^{-16}. A formula for h that balances the rounding error against the secant error for optimum accuracy is:

$$h = 2\sqrt{\varepsilon\left|\frac{f(x)}{f''(x)}\right|}$$

(though not when $f''(x) = 0$), and to employ it will require knowledge of the function.

For single precision the problems are exacerbated because, although x may be a representable floating-point number number, $x + h$ almost certainly will not be. This means that $x + h$ will be changed (by rounding or truncation) to a nearby machine-representable number, with the consequence that $(x + h) - x$ will *not* equal h; the two function evaluations will not be exactly h apart. In this regard, since most decimal fractions are recurring sequences in binary (just as 1/3 is in decimal) a seemingly round step such as $h = 0.1$ will not be a round number in binary; it is $0.000110011001100..._2$ A possible approach is as follows:

```
h := sqrt(eps) * x;

xph := x + h;

dx := xph - x;

slope := (F(xph) - F(x)) / dx;
```

However, with computers, compiler optimization facilities may fail to attend to the details of actual computer arithmetic and instead apply the axioms of mathematics to deduce that *dx* and *h* are the same. With C and similar languages, a directive that *xph* is a volatile variable will prevent this.

Other Methods

Higher-order Methods

Higher-order methods for approximating the derivative, as well as methods for higher derivatives, exist.

Given below is the five-point method for the first derivative (five-point stencil in one dimension):

$$f'(x) = \frac{-f(x+2h)+8f(x+h)-8f(x-h)+f(x-2h)}{12h} + \frac{h^4}{30} f^{(5)}(c),$$

where $c \in [x-2h, x+2h]$.

For other stencil configurations and derivative orders, the Finite Difference Coefficients Calculator is a tool that can be used to generate derivative approximation methods for any stencil with any derivative order (provided a solution exists).

Higher Derivatives

Using Newton's difference quotient,

$$f'(x) = \lim_{h \to 0} \frac{f(x+h)-f(x)}{h},$$

the following can be shown (for n>0):

$$f^{(n)}(x) = \lim_{h \to 0} \frac{1}{h^n} \sum_{k=0}^{n} (-1)^{k+n} \binom{n}{k} f(x+kh).$$

Complex-variable Methods

The classical finite-difference approximations for numerical differentiation are ill-conditioned. However, if *f* is a holomorphic function, real-valued on the real line, which can be evaluated at points in the complex plane near *x*, then there are stable methods. For example, the first derivative can be calculated by the complex-step derivative formula:

$$f'(x) = \Im(f(x+ih))/h + O(h^2), \; i := \sqrt{-1}.$$

This formula can be obtained by Taylor series expansion:

$$f(x+ih) = f(x) + ihf'(x) - h^2 f''(x)/2! - ih^3 f^{(3)}(x)/3! + \cdots.$$

The complex-step derivative formula is only valid for calculating first-order derivatives. A generalization of the above for calculating derivatives of any order employ multicomplex numbers, resulting in multicomplex derivatives.

In general, derivatives of any order can be calculated using Cauchy's integral formula:

$$f^{(n)}(a) = \frac{n!}{2\pi i} \oint_{\gamma} \frac{f(z)}{(z-a)^{n+1}} dz,$$

where the integration is done numerically.

Using complex variables for numerical differentiation was started by Lyness and Moler in 1967. A method based on numerical inversion of a complex Laplace transform was developed by Abate and Dubner. An algorithm that can be used without requiring knowledge about the method or the character of the function was developed by Fornberg.

Differential Quadrature

Differential quadrature is the approximation of derivatives by using weighted sums of function values. The name is in analogy with *quadrature*, meaning numerical integration, where weighted sums are used in methods such as Simpson's method or the Trapezoidal rule. There are various methods for determining the weight coefficients. Differential quadrature is used to solve partial differential equations.

ERROR

In numerical analysis, round-off error is exemplified by the difference between the true value of the irrational number π and the value of rational expressions such as 22/7, 355/113, 3.14, or 3.14159. Truncation error results from ignoring all but a finite number of terms of an infinite series. For example, the exponential function e^x may be expressed as the sum of the infinite series:

$$1 + x + x^2/2 + x^3/6 + L + x^n/n! + ...$$

Stopping the calculation after any finite value of n will give an approximation to the value of e^x that will be in error, but this error can be made as small as desired by making n large enough.

The relative error is the numerical difference divided by the true value; the percentage error is this ratio expressed as a percent. The term random error is sometimes used to distinguish the effects of inherent imprecision from so-called systematic error, which may originate in faulty assumptions or procedures. The methods of mathematical statistics are particularly suited to the estimation and management of random errors.

Types of Error

In the world of math, numerical analysis is well known for focusing on the algorithms used to solve issues in continuous math. The practice is familiar territory for engineers and those who work with physical science, but it's beginning to expand further into liberal arts areas as well. You can see this in astrology, stock portfolio analysis, data analysis and medicine. Part of the application of numerical analysis involves the use of errors. Specific errors are sought out and applied to arrive at mathematical conclusions.

Rounding Off

The round-off error is used because representing every number as a real number isn't possible. So rounding is introduced to adjust for this situation. A round-off error represents the numerical amount between what a figure actually is versus its closest real number value, depending on how the round is applied. For instance, rounding to the nearest whole number means you round up or down to what is the closest whole figure. So if your result is 3.31 then you would round to 3. Rounding the highest amount would be a bit different. In this approach, if your figure is 3.31, your rounding would be to 4. In terms of numerical analysis the round-off error is an attempt to identify what the rounding distance is when it comes up in algorithms. It's also known as a quantization error.

Truncation Error

A truncation error occurs when approximation is involved in numerical analysis. The error factor is related to how much the approximate value is at variance from the actual value in a formula or math result. For example, take the formula of 3 x 3 + 4. The calculation equals 28. Now, break it down and the root is close to 1.99. The truncation error value is therefore equal to 0.01.

Discretization Error

Discretization involves converting or partitioning variables or continuous attributes to nominal attributes, intervals and variables. As a type of truncation error, the discretization error focuses on how much a discrete math problem is not consistent with a continuous math problem.

Numerical Stability

If an error stays at one point in an algorithm and doesn't aggregate further as the calculation continues, then it's considered a numerically stable error. This happens when the error causes only a very small variation in the formula result. If the opposite occurs and the error propagates bigger as the calculation continues, then it is considered numerically unstable.

Error Benefits

Errors are usually regarded as negative, but math errors come in useful in statistics, computer programming, advanced mathematics and much more. Evaluating errors provides significantly useful information, especially when chance and probability is required.

AUTOMATIC COMPUTATIONAL DIFFERENTIATION

In mathematics and computer algebra, automatic differentiation (AD), also called algorithmic differentiation or computational differentiation, is a set of techniques to numerically evaluate the derivative of a function specified by a computer program. AD exploits the fact that every computer program, no matter how complicated, executes a sequence of elementary arithmetic operations (addition, subtraction, multiplication, division, etc.) and elementary functions (exp, log, sin, cos, etc.). By applying the chain rule repeatedly to these operations, derivatives of arbitrary order can be computed automatically, accurately to working precision, and using at most a small constant factor more arithmetic operations than the original program.

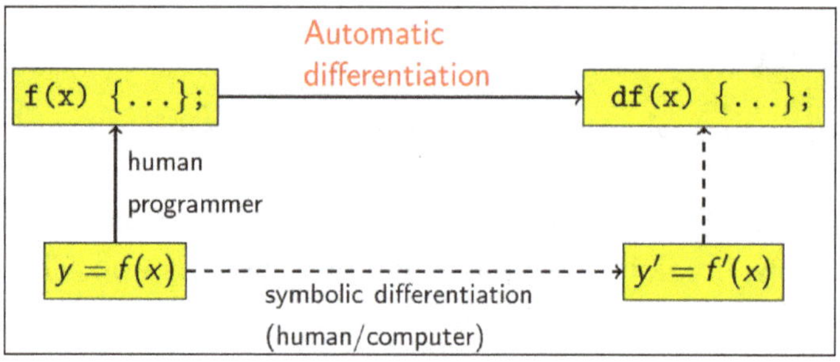

How automatic differentiation relates to symbolic differentiation.

Automatic differentiation is distinct from symbolic differentiation and numerical differentiation (the method of finite differences). Symbolic differentiation can lead to inefficient code and faces the difficulty of converting a computer program into a single expression, while numerical differentiation can introduce round-off errors in the discretization process and cancellation. Both classical methods have problems with calculating higher derivatives, where complexity and errors increase. Finally, both classical methods are slow at computing partial derivatives of a function with respect to *many* inputs, as is needed for gradient-based optimization algorithms. Automatic differentiation solves all of these problems, at the expense of introducing more software dependencies.

Chain Rule, Forward and Reverse Accumulation

Fundamental to AD is the decomposition of differentials provided by the chain rule. For the simple composition:

$$
\begin{aligned}
&y = f(g(h(x))) = f(g(h(w_0))) = f(g(w_1)) = f(w_2) = w_3 \\
&w_0 = x \\
&w_1 = h(w_0) \\
&w_2 = g(w_1) \\
&w_3 = f(w_2) = y
\end{aligned}
$$

The chain rule gives:

$$\frac{dy}{dx} = \frac{dy}{dw_2}\frac{dw_2}{dw_1}\frac{dw_1}{dx}$$

Usually, two distinct modes of AD are presented, forward accumulation (or forward mode) and reverse accumulation (or reverse mode). Forward accumulation specifies that one traverses the chain rule from inside to outside (that is, first compute dw_1 / dx and then dw_2 / dw_1 and at last dy / dw_2), while reverse accumulation has the traversal from outside to inside (first compute dy / dw_2 and then dw_2 / dw_1 and at last dw_1 / dx). More succinctly,

- Forward accumulation computes the recursive relation: $\frac{dw_i}{dx} = \frac{dw_i}{dw_{i-1}}\frac{dw_{i-1}}{dx}$ with $w_3 = y$, and,
- Reverse accumulation computes the recursive relation: $\frac{dy}{dw_i} = \frac{dy}{dw_{i+1}}\frac{dw_{i+1}}{dw_i}$ with $w_0 = x$.

Generally, both forward and reverse accumulation are specific manifestations of applying the operator of program composition, fixing the appropriate one of the two mappings (w,y).

Forward Accumulation

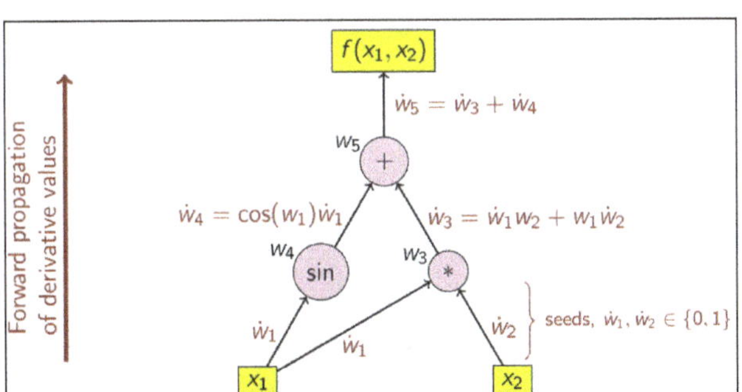

Example of forward accumulation with computational graph.

In forward accumulation AD, one first fixes the independent variable with respect to which differentiation is performed and computes the derivative of each sub-expression recursively. In a pen-and-paper calculation, this involves repeatedly substituting the derivative of the inner functions in the chain rule:

$$\frac{\partial y}{\partial x}=\frac{\partial y}{\partial w_{n-1}}\frac{\partial w_{n-1}}{\partial x}=\frac{\partial y}{\partial w_{n-1}}\left(\frac{\partial w_{n-1}}{\partial w_{n-2}}\frac{\partial w_{n-2}}{\partial x}\right)=\frac{\partial y}{\partial w_{n-1}}\left(\frac{\partial w_{n-1}}{\partial w_{n-2}}\left(\frac{\partial w_{n-2}}{\partial w_{n-3}}\frac{\partial w_{n-3}}{\partial x}\right)\right)=\cdots$$

This can be generalized to multiple variables as a matrix product of Jacobians.

Compared to reverse accumulation, forward accumulation is natural and easy to implement as the flow of derivative information coincides with the order of evaluation. Each variable w is augmented with its derivative $\dot{w}$ (stored as a numerical value, not a symbolic expression),

$$\dot{w}=\frac{\partial w}{\partial x}$$

as denoted by the dot. The derivatives are then computed in sync with the evaluation steps and combined with other derivatives via the chain rule.

As an example, consider the function:

$$\begin{aligned}z &= f(x_1,x_2)\\ &= x_1x_2+\sin x_1\\ &= w_1w_2+\sin w_1\\ &= w_3+w_4\\ &= w_5\end{aligned}$$

For clarity, the individual sub-expressions have been labeled with the variables w_i.

The choice of the independent variable to which differentiation is performed affects the *seed* values $\dot{w}_1$ and $\dot{w}_2$. Given interest in the derivative of this function with respect to x_1, the seed values should be set to:

$$\dot{w}_1=\frac{\partial x_1}{\partial x_1}=1$$

$$\dot{w}_2=\frac{\partial x_2}{\partial x_1}=0$$

With the seed values set, the values propagate using the chain rule as shown. Figure shows a pictorial depiction of this process as a computational graph.

Operations to compute value	Operations to compute derivative
$w_1 = x_1$	$\dot{w}_1 = 1$ (seed)
$w_2 = x_2$	$\dot{w}_2 = 0$ (seed)
$w_3 = w_1 \cdot w_2$	$\dot{w}_3 = w_2 \cdot \dot{w}_1 + w_1 \cdot \dot{w}_2$
$w_4 = \sin w_1$	$\dot{w}_4 = \cos w_1 \cdot \dot{w}_1$
$w_5 = w_3 + w_4$	$\dot{w}_5 = \dot{w}_3 + \dot{w}_4$

To compute the gradient of this example function, which requires the derivatives of f with respect to not only x_1 but also x_2, an *additional* sweep is performed over the computational graph using the seed values $\dot{w}_1 = 0; \dot{w}_2 = 1$.

The computational complexity of one sweep of forward accumulation is proportional to the complexity of the original code.

Forward accumulation is more efficient than reverse accumulation for functions $f: \mathrm{R}^n \rightarrow \mathrm{R}^m$ with $m \gg n$ as only n sweeps are necessary, compared to m sweeps for reverse accumulation.

Reverse Accumulation

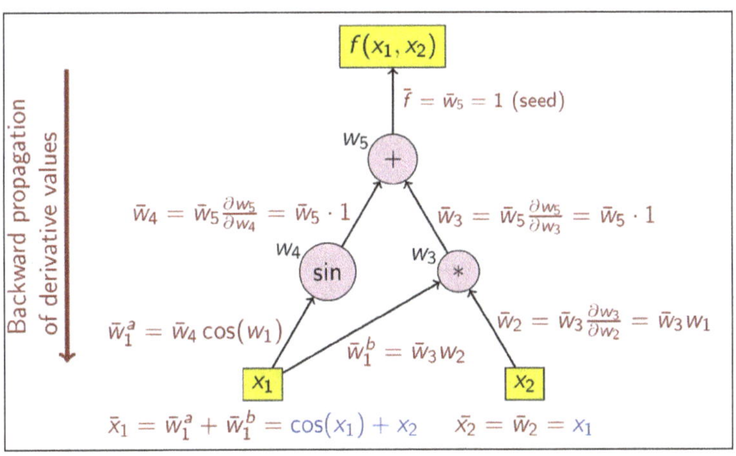

Example of reverse accumulation with computational graph.

In reverse accumulation AD, the *dependent variable* to be differentiated is fixed and the derivative is computed *with respect to* each sub-expression recursively. In a pen-and-paper calculation, the derivative of the *outer* functions is repeatedly substituted in the chain rule:

$$\frac{\partial y}{\partial x} = \frac{\partial y}{\partial w_1}\frac{\partial w_1}{\partial x} = \left(\frac{\partial y}{\partial w_2}\frac{\partial w_2}{\partial w_1}\right)\frac{\partial w_1}{\partial x} = \left(\left(\frac{\partial y}{\partial w_3}\frac{\partial w_3}{\partial w_2}\right)\frac{\partial w_2}{\partial w_1}\right)\frac{\partial w_1}{\partial x} = \cdots$$

In reverse accumulation, the quantity of interest is the *adjoint*, denoted with a bar ($\bar{w}$); it is a derivative of a chosen dependent variable with respect to a subexpression w:

$$\bar{w} = \frac{\partial y}{\partial w}$$

Reverse accumulation traverses the chain rule from outside to inside, or in the case of the computational graph in figure, from top to bottom. The example function is scalar-valued, and thus there is only one seed for the derivative computation, and only one sweep of the computational graph is needed to calculate the (two-component) gradient. This is only half the work when compared to forward accumulation, but reverse accumulation requires the storage of the intermediate variables w_i as well as the instructions that produced them in a data structure known as a Wengert list (or "tape"), which may consume significant memory if the computational graph is large. This can be mitigated to some extent by storing only a subset of the intermediate variables and then reconstructing the necessary work variables by repeating the evaluations, a technique known as rematerialization. Checkpointing is also used to save intermediary states.

The operations to compute the derivative using reverse accumulation are shown in the table (note the reversed order):

Operations to compute derivative
$\bar{w}_5 = 1$ (seed)
$\bar{w}_4 = \bar{w}_5$
$\bar{w}_3 = \bar{w}_5$
$\bar{w}_2 = \bar{w}_3 \cdot w_1$
$\bar{w}_1 = \bar{w}_3 \cdot w_2 + \bar{w}_4 \cdot \cos w_1$

The data flow graph of a computation can be manipulated to calculate the gradient of its original calculation. This is done by adding an adjoint node for each primal node, connected by adjoint edges which parallel the primal edges but flow in the opposite direction. The nodes in the adjoint graph represent multiplication by the derivatives of the functions calculated by the nodes in the primal. For instance, addition in the primal causes fanout in the adjoint; fanout in the primal causes addition in the adjoint; a unary function $y = f(x)$ in the primal causes $\bar{x} = \bar{y} f'(x)$ in the adjoint; etc.

Reverse accumulation is more efficient than forward accumulation for functions $f: \mathrm{R}^n \rightarrow \mathrm{R}^m$ with $m \ll n$ as only m sweeps are necessary, compared to n sweeps for forward accumulation.

Reverse mode AD was first published in 1976 by Seppo Linnainmaa.

Backpropagation of errors in multilayer perceptrons, a technique used in machine learning, is a special case of reverse mode AD.

Beyond Forward and Reverse Accumulation

Forward and reverse accumulation are just two (extreme) ways of traversing the chain rule. The problem of computing a full Jacobian of $f : \mathbb{R}^n \to \mathbb{R}^m$ with a minimum number of arithmetic operations is known as the *optimal Jacobian accumulation* (OJA) problem, which is NP-complete. Central to this proof is the idea that algebraic dependencies may exist between the local partials that label the edges of the graph. In particular, two or more edge labels may be recognized as equal. The complexity of the problem is still open if it is assumed that all edge labels are unique and algebraically independent.

Automatic Differentiation using Dual Numbers

Forward mode automatic differentiation is accomplished by augmenting the algebra of real numbers and obtaining a new arithmetic. An additional component is added to every number to represent the derivative of a function at the number, and all arithmetic operators are extended for the augmented algebra. The augmented algebra is the algebra of dual numbers. This approach was generalized by the theory of operational calculus on programming spaces, through tensor algebra of the dual space.

Replace every number x with the number $x + x'\varepsilon$, where x' is a real number, but ε is an abstract number with the property $\varepsilon^2 = 0$ (an infinitesimal). Using only this, regular arithmetic gives:

$$(x + x'\varepsilon) + (y + y'\varepsilon) = x + y + (x' + y')\varepsilon$$
$$(x + x'\varepsilon) \cdot (y + y'\varepsilon) = xy + xy'\varepsilon + yx'\varepsilon + x'y'\varepsilon^2 = xy + (xy' + yx')\varepsilon$$

and likewise for subtraction and division.

Now, polynomials can be calculated in this augmented arithmetic. If:

$$\begin{aligned}
P(x) &= p_0 + p_1 x + p_2 x^2 + \cdots + p_n x^n, \text{then} \\
P(x + x'\varepsilon) &= p_0 + p_1(x + x'\varepsilon) + \cdots + p_n(x + x'\varepsilon)^n \\
&= p_0 + p_1 x + \cdots + p_n x^n + p_1 x'\varepsilon + 2p_2 x x'\varepsilon + \cdots + n p_n x^{n-1} x'\varepsilon \\
&= P(x) + P^{(1)}(x) x'\varepsilon
\end{aligned}$$

Where $P^{(1)}$ denotes the derivative of P with respect to its first argument, and x', called a *seed*, can be chosen arbitrarily.

The new arithmetic consists of ordered pairs, elements written $\langle x, x' \rangle$, with ordinary arithmetics on the first component, and first order differentiation arithmetic on the second component. Extending the results on polynomials to analytic functions gives a list of the basic arithmetic and some standard functions for the new arithmetic:

$$\langle u, u' \rangle + \langle v, v' \rangle = \langle u + v, u' + v' \rangle$$

$$\langle u,u'\rangle - \langle v,v'\rangle = \langle u-v, u'-v'\rangle$$
$$\langle u,u'\rangle * \langle v,v'\rangle = \langle uv, u'v+uv'\rangle$$
$$\langle u,u'\rangle / \langle v,v'\rangle = \left\langle \frac{u}{v}, \frac{u'v-uv'}{v^2} \right\rangle \; (v \neq 0)$$
$$\sin\langle u,u'\rangle = \langle \sin(u), u'\cos(u)\rangle$$
$$\cos\langle u,u'\rangle = \langle \cos(u), -u'\sin(u)\rangle$$
$$\exp\langle u,u'\rangle = \langle \exp u, u'\exp u\rangle$$
$$\log\langle u,u'\rangle = \langle \log(u), u'/u\rangle \; (u > 0)$$
$$\langle u,u'\rangle^k = \langle u^k, ku^{k-1}u'\rangle \; (u \neq 0)$$
$$|\langle u,u'\rangle| = \langle |u|, u' signu\rangle \; (u \neq 0)$$

and in general for the primitive function g,

$$g(\langle u,u'\rangle, \langle v,v'\rangle) = \langle g(u,v), g_u(u,v)u' + g_v(u,v)v'\rangle$$

where g_u and g_v are the derivatives of g with respect to its first and second arguments, respectively.

When a binary basic arithmetic operation is applied to mixed arguments—the pair $\langle u,u'\rangle$ and the real number c—the real number is first lifted to $\langle c,0\rangle$. The derivative of a function $f:\mathbb{R}\to\mathbb{R}$ at the point is now found by calculating $f(\langle x_0,1\rangle)$ using the above arithmetic, which gives $\langle f(x_0), f'(x_0)\rangle$ as the result.

Vector Arguments and Functions

Multivariate functions can be handled with the same efficiency and mechanisms as univariate functions by adopting a directional derivative operator. That is, if it is sufficient to compute $y' = \nabla f(x)\cdot x'$, the directional derivative $y' \in \mathbb{R}^m$ of $f:\mathbb{R}^n \to \mathbb{R}^m$ at $x \in \mathbb{R}^n$ in the direction $x \in \mathbb{R}^n$, this may be calculated as $(\langle y_1,y_1'\rangle,\ldots,\langle y_m,y_m'\rangle) = f(\langle x_1,x_1'\rangle,\ldots,\langle x_n,x_n'\rangle)$ using the same arithmetic. If all the elements of ∇f are desired, then n function evaluations are required. Note that in many optimization applications, the directional derivative is indeed sufficient.

High Order and many Variables

The above arithmetic can be generalized to calculate second order and higher derivatives of multivariate functions. However, the arithmetic rules quickly grow complicated: complexity is quadratic in the highest derivative degree. Instead, truncated Taylor polynomial algebra can be used. The resulting arithmetic, defined on generalized dual numbers, allows efficient computation using functions as if they were a data type. Once the Taylor polynomial of a function is known, the derivatives are easily extracted. A rigorous, general formulation is achieved through the tensor series expansion using operational calculus on programming spaces.

Operational Calculus on Programming Spaces

Operational calculus on programming spaces provides differentiable programming with formal semantics through an algebra of higher-order constructs. It can thus be used to express the concepts underlying automatic differentiation.

Differentiable Programming Space

A differentiable programming space $\mathcal{P}_0$ is any subspace of $\mathcal{F}_0 : \mathcal{V} \to \mathcal{V}$ such that:

$$\partial \mathcal{P}_0 \subset \mathcal{P}_0 \otimes T(\mathcal{V}^*),$$

where $T(\mathcal{V}^*)$ is the tensor algebra of the dual space $\mathcal{V}^*$. When all elements of $\mathcal{P}_0$ are analytic, we call $\mathcal{P}_0$ an analytic programming space.

Theorem: Any differentiable programming space $\mathcal{P}_0$ is an infinitely differentiable programming space, meaning that,

$$\partial^k \mathcal{P}_0 \subset \mathcal{P}_0 \otimes T(\mathcal{V}^*)$$

for any $k \in \mathbb{N}$. If all elements of $\mathcal{P}_0$ are analytic, than so are the elements of $\mathcal{P}_n$.

Definition: Let $\mathcal{P}_0$ be a differentiable programming space. The space $\mathcal{P}_n < \mathcal{F}_n$: $\mathcal{V} \to \mathcal{V} \otimes T(\mathcal{V}^*)$ spanned by $\mathcal{D}^n \mathcal{P}_0$ over K, where $\mathcal{D}^n = \{\partial^k \mid 0 \le k \le n\}$, is called a differentiable programming space of order n.

Corollary: A differentiable programming space of order $n, \mathcal{P}_n : \mathcal{V} \to \mathcal{V} \otimes T(\mathcal{V}^*)$, can be embedded into the tensor product of the function space $\mathcal{P}_0 : \mathcal{V} \to \mathcal{V}$ and the subspace $T_n(\mathcal{V}^*)$ of the tensor algebra of the dual of the virtual space $\mathcal{P}$. By taking the limit as $n \to \infty$, we consider,

$$\mathcal{P}_\infty < \mathcal{P}_0 \otimes T(\mathcal{V}^*),$$

where,

$$T(\mathcal{V}^*) = \prod_{k=0}^{\infty} (\mathcal{V}^*)^{\otimes k}$$

is the tensor series algebra, the algebra of the infinite formal tensor series, which is a completion of the tensor algebra $T(\mathcal{V}^*)$ in suitable topology.

This means that calculation of derivatives of the map $P:\mathcal{V}\to\mathcal{V}$, can be calculated with only one mapping τ. The operator τ_n as a direct sum of operators:

$$\tau_n = 1+\partial+\partial^2+\cdots+\partial^n$$

The image $\tau_k P(\mathrm{x})$ is a multitensor of order k, which is a direct sum of the maps value and all derivatives of order $n\le k$, all evaluated at the point x:

$$\tau_k P(\mathrm{x}) = P(\mathrm{x})+\partial_{\mathrm{x}}P(\mathrm{x})+\partial_{\mathrm{x}}^2 P(\mathrm{x})+\cdots+\partial_{\mathrm{x}}^k P(\mathrm{x}).$$

The operator τ_n satisfies the recursive relation:

$$\tau_{k+1} = 1+\partial\tau_k,$$

that can be used to recursively construct programming spaces of arbitrary order. Only explicit knowledge of $\tau:\mathcal{P}_0\to\mathcal{P}_1$ is required for the construction of $\mathcal{P}_n$ from $\mathcal{P}_1$, which is evident from the above theorem.

Virtual Tensor Machine

An abstract virtual machine is capable of constructing and implementing the theory. Such a machine provides a framework for analytic study of algorithmic procedures through algebraic means.

Claim: The tuple $(\mathcal{V},\mathcal{P}_0)$ and the belonging tensor series algebra are sufficient conditions for the existence and construction of infinitely differentiable programming spaces $\mathcal{P}_\infty$, through linear combinations of elements of $\mathcal{P}_0\otimes T(\mathcal{V}^*)$.

This claim allows a simple definition of such a machine.

Definition (Virtual tensor machine): The tuple $M=\langle\mathcal{V},\mathcal{P}_0\rangle$ is a virtual tensor machine, where,

- $\mathcal{V}$ is a finite dimensional vector space.
- $\mathcal{V}\otimes T(\mathcal{V}^*)$ is the virtual memory space.
- $\mathcal{P}_0$ is an analytic programming space over $\mathcal{V}$.

Tensor Series Expansion

Expansion into a series offers valuable insights into programs through methods of analysis.

There exists a space spanned by the set $\mathcal{D}^n = \{\partial^k \mid 0 \le k \le n\}$ over a field K. Thus, the expression:

$$e^{h\partial} = \sum_{n=0}^{\infty} \frac{(h\partial)^n}{n!}$$

is well defined. The operator $e^{h\partial}$ is a mapping between function spaces:

$$e^{h\partial} : \mathcal{P} \to \mathcal{P}_\infty.$$

It also defines a map:

$$e^{h\partial} : \mathcal{P} \times \mathcal{V} \to \mathcal{V} \otimes \mathcal{T}(\mathcal{V}^*),$$

by taking the image of the map $e^{h\partial}(P)$ at a certain point $\mathrm{v} \in \mathcal{V}$.

A map can be constructed from the space of programs to the space of polynomials. The space of multivariate polynomials $\mathcal{V} \to K$ is isomorphic to symmetric algebra $\mathcal{S}(\mathcal{V}^*)$, which is in turn a quotient of tensor algebra $T(\mathcal{V}^*)$. To any element of $\mathcal{V} \otimes T(\mathcal{V}^*)$ the corresponding element of $\mathcal{V} \otimes S(\mathcal{V}^{i*})$ can be attached, namely a polynomial map $\mathcal{V} \to \mathcal{V}$. Thus, the completion of the symmetric algebra $\mathcal{S}(\mathcal{V}^*)$ is the *formal power series* $\mathcal{S}(\mathcal{V}^*)$, which is in turn isomorphic to a quotient of *tensor series algebra* $\mathcal{T}(\mathcal{V}^*)$, arriving at

$$e^{h\partial} : \mathcal{P} \times \mathcal{V} \to \mathcal{V} \otimes \mathcal{S}(\mathcal{V}^{i*})$$

For any element $\mathrm{v}_0 \in \mathcal{V}$, the expression $e^{h\partial}(\cdot, \mathrm{v}_0)$ is a map $\mathcal{P} \to \mathcal{V} \otimes \mathcal{S}(\mathcal{V}^*)$,, mapping a program to a *formal power series*. The correspondence between multi-tensors in $\mathcal{V} \otimes T(\mathcal{V}^*)$ and polynomial maps $\mathcal{V} \to \mathcal{V}$ can be given by multiple contractions for all possible indices.

Theorem: For a program $P \in \mathcal{P}$ the expansion into an infinite tensor series at the point $\mathrm{v}_0 \in \mathcal{V}$ is expressed by multiple contractions,

$$P(\mathrm{v}_0 + h\mathrm{v}) = \left((e^{h\partial} P)(\mathrm{v}_0)\right) \cdot \mathrm{v} = \sum_{n=0}^{\infty} \frac{h^n}{n!} \partial^n P(\mathrm{v}_0) \cdot (\mathrm{v}^{\otimes n})$$

Evaluated at $h = 1$, the operator is a generalization of the shift operator widely used in physics. For a specific $v_0 \in \mathcal{V}$ it is here on denoted by:

$$e^{\partial}\big|_{v_0} : \mathcal{P} \to \mathcal{V} \otimes \mathcal{T}(\mathcal{V}^*).$$

When the choice of $v_0 \in \mathcal{V}$ is arbitrary, it is omitted from expressions for brevity. Others adopted similar approaches.

Operator of Program Composition

Theory offers a generalization of both forward and reverse mode of automatic differentiation to arbitrary order, under a single invariant operator in the theory. This condenses complex notions to simple expressions allowing meaningful manipulations before being applied to a particular programming space.

Theorem: Composition of maps $\mathcal{P}$ is expressed as,

$$e^{h\partial}(f \circ g) = \exp(\partial_f e^{h\partial_g})(g, f)$$

where $\exp(\partial_f e^{h\partial_g}) : \mathcal{P} \times \mathcal{P} \to \mathcal{P}_\infty$ is an operator on pairs of maps (g, f), where ∂_g is applied to g and ∂_f to f.

Both forward and reverse mode (generalized to arbitrary order) are obtainable using this operator, by fixing the appropriate one of the two maps. This generalizes both concepts under a single operator in the theory. For example, by considering projections of the operator onto the space spanned by $\mathcal{D} = \{1, \partial\}$, and fixing the second map g, we retrieve the basic first order forward mode of automatic differentiation, or reverse mode, by fixing f.

Thus the operator alleviates the need for explicit implementation of the higher order chain rule, as it is encoded in the structure of the operator itself, which can be efficiently implemented by manipulating its generating map.

Order Reduction for Nested Applications

It is useful to be able to use the k-th derivative of a program $P \in \mathcal{P}$ as part of a different differentiable program P_1. As such, we must be able to treat the derivative itself as a differentiable program $P'^k \in \mathcal{P}$, while only coding the original program P.

Theorem: There exists a reduction of order map $\phi : \mathcal{P}_n \to \mathcal{P}_{n-1}$ satisfying,

$$\forall P_1 \in \mathcal{P}_0 \quad \exists P_2 \in \mathcal{P}_0 \qquad \phi^k \circ e_n^\partial (P_1) = e_{n-k}^\partial (P_2)$$

for each $n \geq 1$, where e_n^∂ is the projection of the operator e^∂ onto the set $\mathcal{D}^n = \{\partial^k \mid 0 \leq k \leq n\}$.

By the above theorem, n-differentiable k-th derivatives of a program $P \in \mathcal{P}_0$ can be extracted by:

$$^{n}P^{k'} = \phi^k \circ e_{n+k}^\partial (P) \in \mathcal{P}_n.$$

Thus, we gained the ability of writing a differentiable program acting on derivatives of another program.

Implementation

Forward-mode AD is implemented by a nonstandard interpretation of the program in which real numbers are replaced by dual numbers, constants are lifted to dual numbers with a zero epsilon coefficient, and the numeric primitives are lifted to operate on dual numbers. This nonstandard interpretation is generally implemented using one of two strategies: source code transformation or operator overloading.

Source Code Transformation (SCT)

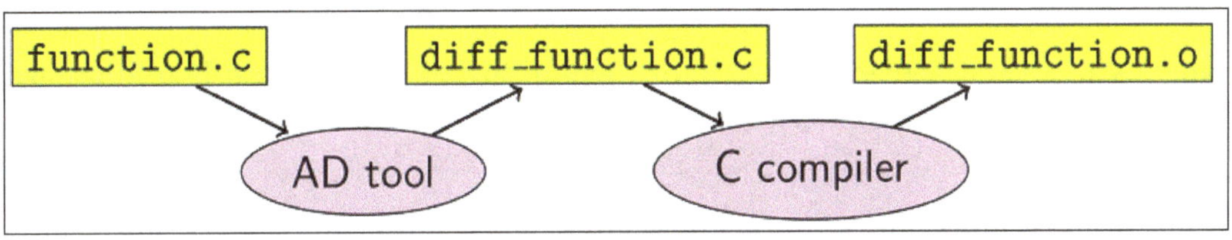

Example of how source code transformation could work.

The source code for a function is replaced by an automatically generated source code that includes statements for calculating the derivatives interleaved with the original instructions.

Source code transformation can be implemented for all programming languages, and it is also easier for the compiler to do compile time optimizations. However, the implementation of the AD tool itself is more difficult.

Operator Overloading (OO)

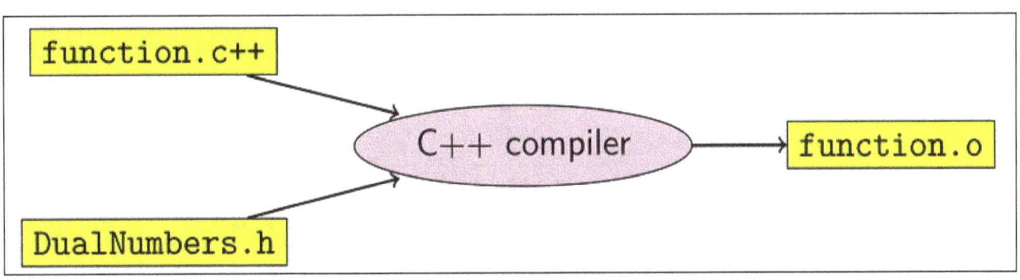

Example of how operator overloading could work.

Operator overloading is a possibility for source code written in a language supporting it. Objects for real numbers and elementary mathematical operations must be overloaded to cater for the augmented arithmetic depicted before. This requires no change in the form or sequence of operations in the original source code for the function to be differentiated, but often requires changes in basic data types for numbers and vectors to support overloading and often also involves the insertion of special flagging operations.

Operator overloading for forward accumulation is easy to implement, and also possible for reverse accumulation. However, current compilers lag behind in optimizing the code when compared to forward accumulation.

Operator overloading, for both forward and reverse accumulation, can be well-suited to applications where the objects are vectors of real numbers rather than scalars. This is because the tape then comprises vector operations; this can facilitate computationally efficient implementations where each vector operation performs many scalar operations. Vector adjoint algorithmic differentiation (vector AAD) techniques may be used, for example, to differentiate values calculated by Monte-Carlo simulation.

Examples of operator-overloading implementations of automatic differentiation in C++ are the Adept and Stan libraries.

NUMERICAL INTEGRATION

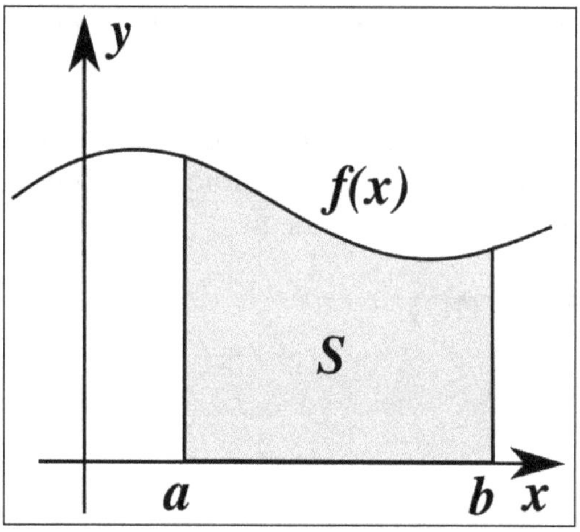

Numerical integration is used to calculate a numerical approximation for the value S, the area under the curve defined by $f(x)$.

In analysis, numerical integration comprises a broad family of algorithms for calculating the numerical value of a definite integral, and by extension, the term is also sometimes used to describe the numerical solution of differential equations. Here, we will focus on calculation of definite integrals. The term numerical quadrature (often abbreviated to *quadrature*) is more or less a synonym for *numerical integration*, especially as applied to one-dimensional integrals.

The basic problem in numerical integration is to compute an approximate solution to a definite integral:

$$\int_a^b f(x)dx$$

to a given degree of accuracy. If $f(x)$ is a smooth function integrated over a small number of dimensions, and the domain of integration is bounded, there are many methods for approximating the integral to the desired precision.

Reasons for Numerical Integration

There are several reasons for carrying out numerical integration.

- The integrand $f(x)$ may be known only at certain points, such as obtained by sampling. Some embedded systems and other computer applications may need numerical integration for this reason.
- A formula for the integrand may be known, but it may be difficult or impossible to find an antiderivative that is an elementary function. An example of such an integrand is $f(x) = \exp(-x^2)$, the antiderivative of which (the error function, times a constant) cannot be written in elementary form.
- It may be possible to find an antiderivative symbolically, but it may be easier to compute a numerical approximation than to compute the antiderivative. That may be the case if the antiderivative is given as an infinite series or product, or if its evaluation requires a special function that is not available.

Methods for One-dimensional Integrals

Numerical integration methods can generally be described as combining evaluations of the integrand to get an approximation to the integral. The integrand is evaluated at a finite set of points called integration points and a weighted sum of these values is used to approximate the integral. The integration points and weights depend on the specific method used and the accuracy required from the approximation.

An important part of the analysis of any numerical integration method is to study the behavior of the approximation error as a function of the number of integrand evaluations. A method that yields a small error for a small number of evaluations is usually considered superior. Reducing the number of evaluations of the integrand reduces the number of arithmetic operations involved, and therefore reduces the total round-off error. Also, each evaluation takes time, and the integrand may be arbitrarily complicated.

A 'brute force' kind of numerical integration can be done, if the integrand is reasonably well-behaved (i.e. piecewise continuous and of bounded variation), by evaluating the integrand with very small increments.

Quadrature Rules based on Interpolating Functions

A large class of quadrature rules can be derived by constructing interpolating functions that are easy to integrate. Typically these interpolating functions are polynomials. In practice, since polynomials of very high degree tend to oscillate wildly, only polynomials of low degree are used, typically linear and quadratic.

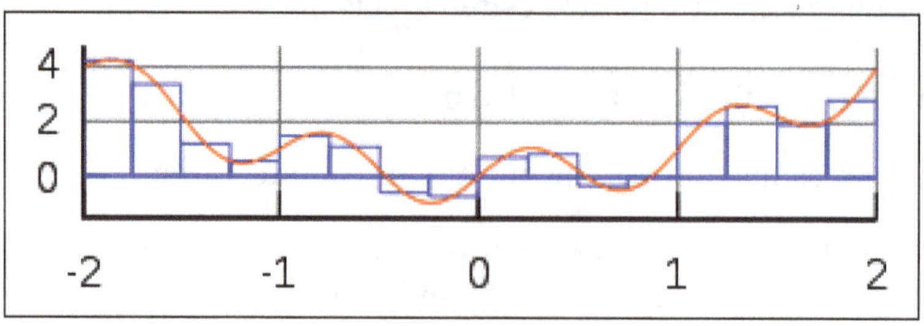

Illustration of the rectangle rule.

The simplest method of this type is to let the interpolating function be a constant function (a polynomial of degree zero) that passes through the point $\left(\frac{a+b}{2}, f\left(\frac{a+b}{2}\right)\right)$. This is called the *midpoint rule* or *rectangle rule:*

$$\int_a^b f(x)dx \approx (b-a)f\left(\frac{a+b}{2}\right).$$

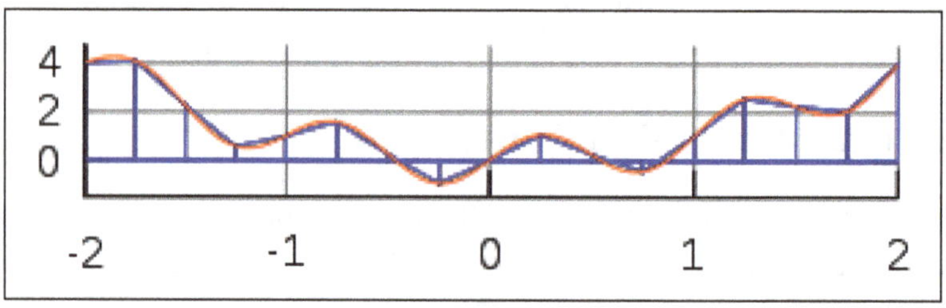

Illustration of the trapezoidal rule.

The interpolating function may be a straight line (an affine function, i.e. a polynomial of degree 1) passing through the points $(a, f(a))$ and $(b, f(b))$. This is called the *trapezoidal rule.*

$$\int_a^b f(x)dx \approx (b-a)\left(\frac{f(a)+f(b)}{2}\right).$$

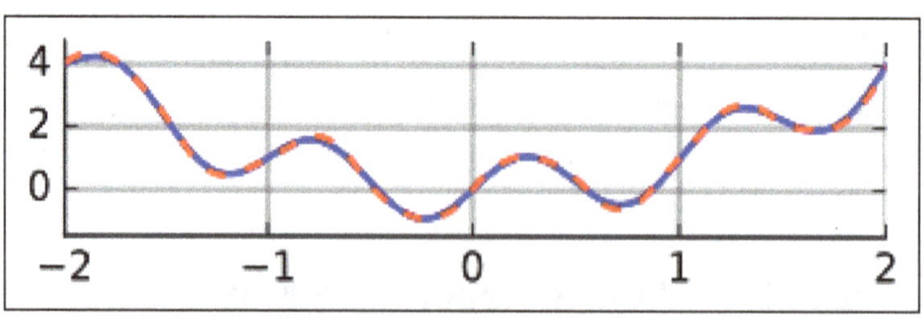

Illustration of Simpson's rule.

For either one of these rules, we can make a more accurate approximation by breaking up the interval [a,b] into some number n of subintervals, computing an approximation for each subinterval, then adding up all the results. This is called a *composite rule*, *extended rule*, or *iterated rule*. For example, the composite trapezoidal rule can be stated as:

$$\int_a^b f(x)dx \approx \frac{b-a}{n}\left(\frac{f(a)}{2}+\sum_{k=1}^{n-1}\left(f\left(a+k\frac{b-a}{n}\right)\right)+\frac{f(b)}{2}\right),$$

where the subintervals have the form $[a+kh, a+(k+1)h] \subset [a,b]$, with $h=\frac{b-a}{n}$ and $k=0,\ldots,n-1$. Here we used subintervals of the same length h but one could also use intervals of varying length $(h_k)_k$.

Interpolation with polynomials evaluated at equally spaced points in [a,b] yields the Newton–Cotes formulas, of which the rectangle rule and the trapezoidal rule are examples. Simpson's rule, which is based on a polynomial of order 2, is also a Newton–Cotes formula.

Quadrature rules with equally spaced points have the very convenient property of *nesting*. The corresponding rule with each interval subdivided includes all the current points, so those integrand values can be re-used.

If we allow the intervals between interpolation points to vary, we find another group of quadrature formulas, such as the Gaussian quadrature formulas. A Gaussian quadrature rule is typically more accurate than a Newton–Cotes rule, which requires the same number of function evaluations, if the integrand is smooth (i.e., if it is sufficiently differentiable). Other quadrature methods with varying intervals include Clenshaw–Curtis quadrature (also called Fejér quadrature) methods, which do nest.

Gaussian quadrature rules do not nest, but the related Gauss–Kronrod quadrature formulas do.

Generalized Midpoint Rule Formula

A generalized midpoint rule formula is given by:

$$\int_0^1 f(x)dx = \sum_{m=1}^{M}\sum_{n=0}^{\infty}\frac{(-1)^n+1}{(2M)^{n+1}(n+1)!}f^{(n)}(x)\Big|_{x=\frac{m-1/2}{M}}$$

or,

$$\int_0^1 f(x)dx = \lim_{M\to\infty}\sum_{m=1}^{M}\sum_{n=0}^{N}\frac{(-1)^n+1}{(2M)^{n+1}(n+1)!}f^{(n)}(x)\Big|_{x=\frac{m-1/2}{M}},$$

where $f^{(n)}(x)$ denotes n-th derivative. For example, substituting $M=1$ and:

$$f(x)=\frac{\theta}{1+\theta^2x^2}$$

in the generalized midpoint rule formula, we obtain an equation of the inverse tangent:

$$\tan^{-1}(\theta)=i\sum_{n=1}^{\infty}\frac{1}{2n-1}\left(\frac{1}{(1+2i/\theta)^{2n-1}}-\frac{1}{(1-2i/\theta)^{2n-1}}\right)=2\sum_{n=1}^{\infty}\frac{1}{2n-1}\frac{a_n(\theta)}{a_n^2(\theta)+b_n^2(\theta)},$$

where $i=\sqrt{-1}$ is imaginary unit and,

$$a_1(\theta)=\frac{2}{\theta},$$
$$b_1(\theta)=1,$$
$$a_n(\theta)=\left(1-\frac{4}{\theta^2}\right)a_{n-1}(\theta)+\frac{4}{\theta}b_{n-1}(\theta),$$
$$b_n(\theta)=\left(1-\frac{4}{\theta^2}\right)b_{n-1}(\theta)-\frac{4}{\theta}a_{n-1}(\theta).$$

Since at each odd n the numerator of the integrand becomes $(-1)^n+1=0$, the generalized midpoint rule formula can be reorganized as:

$$\int_0^1 f(x)dx=2\sum_{m=1}^{M}\sum_{n=0}^{\infty}\frac{1}{(2M)^{2n+1}(2n+1)!}f^{(2n)}(x)\Big|_{x=\frac{m-1/2}{M}}.$$

The following example of Mathematica code generates the plot showing difference between inverse tangent and its approximation truncated at $M=5$ and $N=10$:

```
f[theta_, x_] := theta/(1 + theta^2*x^2);

aTan[theta_, M_, nMax_] :=

    2*Sum[(Function[x, Evaluate[D[f[theta, x], {x, 2*n}]]][(m - 1/2)/

       M])/((2*n + 1)!*(2*M)^(2*n + 1)), {m, 1, M}, {n, 0, nMax}];


Plot[{ArcTan[theta] - aTan[theta, 5, 10]}, {theta, -Pi, Pi},

 PlotRange -> All]
```

For a function $g(t)$ defined over interval (a,b), its integral is:

$$\int_a^b g(t)dt=\int_a^{b-a} g(\tau+a)d\tau=(b-a)\int_0^1 g((b-a)x+a)dx.$$

Therefore, we can apply the generalized midpoint integration formula above by assuming that $f(x)=(b-a)g((b-a)x+a)$.

Adaptive Algorithms

If $f(x)$ does not have many derivatives at all points, or if the derivatives become large, then Gaussian quadrature is often insufficient. In this case, an algorithm similar to the following will perform better:

```
def calculate_definite_integral_of_f(f, initial_step_size):

    """

    This algorithm calculates the definite integral of a function

    from 0 to 1, adaptively, by choosing smaller steps near

    problematic points.

    """

    x = 0.0

    h = initial_step_size

    accumulator = 0.0

    while x < 1.0:

        if x + h > 1.0:

            h = 1.0 - x  # At end of unit interval, adjust last step
to end at 1.

        if error_too_big_in_quadrature_of_f_over_range(f, [x, x + h]):

            h = make_h_smaller(h)

        else:

            accumulator += quadrature_of_f_over_range(f, [x, x + h])

            x += h

            if error_too_small_in_quadrature_of_over_range(f, [x, x +
h]):

                h = make_h_larger(h)  # Avoid wasting time on tiny steps.

    return accumulator
```

Some details of the algorithm require careful thought. For many cases, estimating the error from quadrature over an interval for a function $f(x)$ isn't obvious. One popular solution is to use two different rules of quadrature, and use their difference as an estimate of the error from quadrature. The other problem is deciding what "too large" or "very small" signify. A *local* criterion for "too large" is that the quadrature error should not be larger than $t \cdot h$ where t, a real number, is the tolerance we wish to set for global error. Then again, if h is already tiny, it may not be worthwhile to make it even smaller even if the quadrature error is apparently large. A *global* criterion is that the sum of errors on all the intervals should be less than t. This type of error analysis is usually called "a posteriori" since we compute the error after having computed the approximation.

Extrapolation Methods

The accuracy of a quadrature rule of the Newton-Cotes type is generally a function of the number of evaluation points. The result is usually more accurate as the number of evaluation points increases, or, equivalently, as the width of the step size between the points decreases. It is natural to ask what the result would be if the step size were allowed to approach zero. This can be answered by extrapolating the result from two or more nonzero step sizes, using series acceleration methods such as Richardson extrapolation. The extrapolation function may be a polynomial or rational function.

Conservative a Priori Error Estimation

Let f have a bounded first derivative over $[a,b]$, i.e. $f \in C^1([a,b])$. The mean value theorem for f, where $x \in [a,b)$, gives:

$$(x-a)f'(\xi_x) = f(x) - f(a),$$

for some $\xi_x \in (a,x]$ depending on x.

If we integrate in x from a to b on both sides and take the absolute values, we obtain:

$$\left|\int_a^b f(x)dx - (b-a)f(a)\right| = \left|\int_a^b (x-a)f'(v_x)dx\right|.$$

We can further approximate the integral on the right-hand side by bringing the absolute value into the integrand, and replacing the term in f' by an upper bound:

$$\left|\int_a^b f(x)dx - (b-a)f(a)\right| \le \frac{(b-a)^2}{2} \sup_{a \le x \le b} |f'(x)|,$$

where the supremum was used to approximate.

Hence, if we approximate the integral $\int_a^b f(x)dx$ by the quadrature rule $(b-a)f(a)$ our error is no greater than the right hand side of 1. We can convert this into an error analysis for the Riemann sum (*), giving an upper bound of:

$$\frac{n^{-1}}{2} \sup_{0 \le x \le 1} |f'(x)|$$

for the error term of that particular approximation. (Note that this is precisely the error we calculated for the example $f(x) = x$.) Using more derivatives, and by tweaking the quadrature, we can do a similar error analysis using a Taylor series (using a partial sum with remainder term) for *f*. This error analysis gives a strict upper bound on the error, if the derivatives of *f* are available.

This integration method can be combined with interval arithmetic to produce computer proofs and *verified* calculations.

Integrals over Infinite Intervals

Several methods exist for approximate integration over unbounded intervals. The standard technique involves specially derived quadrature rules, such as Gauss-Hermite quadrature for integrals on the whole real line and Gauss-Laguerre quadrature for integrals on the positive reals. Monte Carlo methods can also be used, or a change of variables to a finite interval; e.g., for the whole line one could use:

$$\int_{-\infty}^{\infty} f(x)dx \quad \int_{-}^{+} f\left(\frac{t}{1-t^2}\right)\frac{t}{(1-t^2)^2}dt,$$

and for semi-infinite intervals one could use:

$$\int_{a}^{\infty} f(x)dx = \int_{0}^{1} f\left(a+\frac{t}{1-t}\right)\frac{dt}{(1-t)^2},$$

$$\int_{-\infty}^{a} f(x)dx = \int_{0}^{1} f\left(a-\frac{1-t}{t}\right)\frac{dt}{t^2},$$

as possible transformations.

Multidimensional Integrals

To compute integrals in multiple dimensions, one approach is to phrase the multiple integral as repeated one-dimensional integrals by applying Fubini's theorem (the tensor product rule). This approach requires the function evaluations to grow exponentially as the number of dimensions increases. Three methods are known to overcome this so-called *curse of dimensionality*.

A great many additional techniques for forming multidimensional cubature integration rules for a variety of weighting functions are given in the monograph by Stroud.

Monte Carlo

Monte Carlo methods and quasi-Monte Carlo methods are easy to apply to multi-dimensional integrals. They may yield greater accuracy for the same number of function evaluations than repeated integrations using one-dimensional methods.

A large class of useful Monte Carlo methods are the so-called Markov chain Monte Carlo algorithms, which include the Metropolis-Hastings algorithm and Gibbs sampling.

Sparse Grids

Sparse grids were originally developed by Smolyak for the quadrature of high-dimensional functions. The method is always based on a one-dimensional quadrature rule, but performs a more sophisticated combination of univariate results. However, whereas the tensor product rule guarantees that the weights of all of the cubature points will be positive if the weights of the quadrature points were positive, Smolyak's rule does not guarantee that the weights will all be positive.

Bayesian Quadrature

Bayesian Quadrature is a statistical approach to the numerical problem of computing integrals and falls under the field of probabilistic numerics. It can provide a full handling of the uncertainty over the solution of the integral expressed as a Gaussian Process posterior variance. It is also known to provide very fast convergence rates which can be up to exponential in the number of quadrature points n.

Connection with Differential Equations

The problem of evaluating the integral:

$$F(x)=\int_a^x f(u)du$$

can be reduced to an initial value problem for an ordinary differential equation by applying the first part of the fundamental theorem of calculus. By differentiating both sides of the above with respect to the argument x, it is seen that the function F satisfies:

$$\frac{dF(x)}{dx}=f(x), \quad F(a)=0.$$

Methods developed for ordinary differential equations, such as Runge–Kutta methods, can be applied to the restated problem and thus be used to evaluate the integral. For instance, the standard fourth-order Runge–Kutta method applied to the differential equation yields Simpson's rule.

The differential equation $F'(x) = f(x)$ has a special form: the right-hand side contains only the independent variable (here x) and not the dependent variable (here F). This simplifies the theory and algorithms considerably. The problem of evaluating integrals is thus best studied in its own right.

Adaptive Quadrature

Adaptive quadrature is a numerical integration method in which the integral of a function *f*(*x*) is approximated using static quadrature rules on adaptively refined subintervals of the integration domain. Generally, adaptive algorithms are just as efficient and effective as traditional algorithms for "well behaved" integrands, but are also effective for "badly behaved" integrands for which traditional algorithms fail.

General Scheme

Adaptive quadrature follows the general scheme:

```
procedure integrate ( f, a, b, τ )
```

$$Q \approx \int_a^b f(x)dx$$

$$\varepsilon \approx \left| Q - \int_a^b f(x)dx \right|$$

```
    if ε>τ then
        m = (a + b) / 2
        Q = integrate(f, a, m, τ/2) + integrate(f, m, b, τ/2)
    endif
    return Q
```

An approximation Q to the integral of *f*(*x*) over the interval [a,b] is computed (line 2), as well as an error estimate ε (line 3). If the estimated error is larger than the required tolerance τ (line 4), the interval is subdivided (line 5) and the quadrature is applied on both halves separately (line 6). Either the initial estimate or the sum of the recursively computed halves is returned (line 7).

The important components are the quadrature rule itself:

$$Q \approx \int_a^b f(x)dx,$$

the error estimator:

$$\varepsilon \approx \left| Q - \int_a^b f(x)dx \right|,$$

and the logic for deciding which interval to subdivide, and when to terminate.

Basic Rules

The quadrature rules generally have the form:

$$Q_n = \sum_{i=0}^{n} w_i f(x_i) \approx \int_a^b f(x)dx$$

where the nodes x_i and weights w_i are generally precomputed.

In the simplest case, Newton–Cotes formulas of even degree are used, where the nodes are evenly spaced in the interval:

$$x_i = a + \frac{i}{n}(b-a).$$

When such rules are used, the points at which $f(x)$ has been evaluated can be re-used upon recursion:

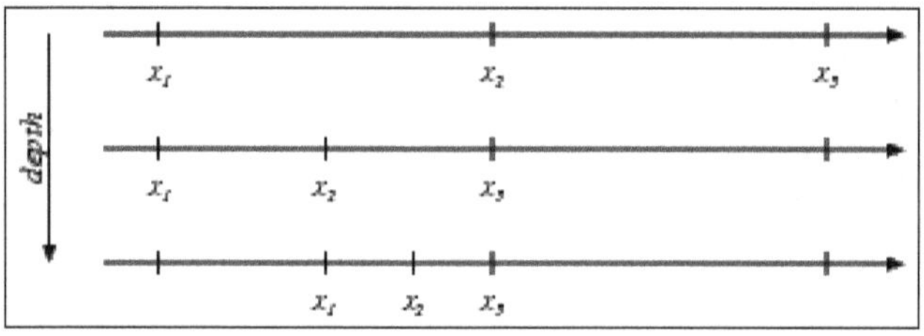

A similar strategy is used with Clenshaw–Curtis quadrature, where the nodes are chosen as:

$$x_i = \cos\left(\frac{2i}{n}\pi\right).$$

Or, when Fejér quadrature is used,

$$x_i = \cos\left(\frac{2(i+0.5)}{n+1}\pi\right).$$

Other quadrature rules, such as Gaussian quadrature or Gauss-Kronrod quadrature, may also be used.

An algorithm may elect to use different quadrature methods on different subintervals, for example using a high-order method only where the integrand is smooth.

Error Estimation

Some quadrature algorithms generate a sequence of results which should approach the correct value. Otherwise one can use a "null rule" which has the form of the above quadrature rule, but whose value would be zero for a simple integrand (for example, if the integrand were a polynomial of the appropriate degree).

Subdivision Logic

"Local" adaptive quadrature makes the acceptable error for a given interval proportional to the length of that interval. This criterion can be difficult to satisfy if the integrands are badly behaved at only a few points, for example with a few step discontinuities. Alternatively, one could require only that the sum of the errors on each of the subintervals be less than the user's requirement. This would be "global" adaptive quadrature. Global adaptive quadrature can be more efficient (using fewer evaluations of the integrand) but is generally more complex to program and may require more working space to record information on the current set of intervals.

Gaussian Quadrature

In numerical analysis, a quadrature rule is an approximation of the definite integral of a function, usually stated as a weighted sum of function values at specified points within the domain of integration. An n-point Gaussian quadrature rule, named after Carl Friedrich Gauss, is a quadrature rule constructed to yield an exact result for polynomials of degree $2n - 1$ or less by a suitable choice of the nodes x_i and weights w_i for $i = 1, ..., n$. Modern formulation using orthogonal polynomials was developed by Carl Gustav Jacobi 1826. The most common domain of integration for such a rule is taken as $[-1, 1]$, so the rule is stated as:

$$\int_{-1}^{1} f(x)dx \approx \sum_{i=1}^{n} w_i f(x_i),$$

which is exact for polynomials of degree $2n - 1$ or less. This exact rule is known as the Gauss-Legendre quadrature rule. The quadrature rule will only be an accurate approximation to the integral above if $f(x)$ is well-approximated by a polynomial of degree $2n - 1$ or less on $[-1, 1]$.

The Gauss-Legendre quadrature rule is not typically used for integrable functions with endpoint singularities. Instead, if the integrand can be written as:

$$f(x) = (1-x)^{\alpha}(1+x)^{\beta} g(x), \quad \alpha, \beta > -1,$$

where $g(x)$ is well-approximated by a low-degree polynomial, then alternative nodes x_i' and weights w_i' will usually give more accurate quadrature rules. These are known as Gauss-Jacobi quadrature rules, i.e.,

$$\int_{-1}^{1} f(x)dx = \int_{-1}^{1} (1-x)^{\alpha} (1+x)^{\beta} g(x)dx \approx \sum_{i=1}^{n} w_i' g(x_i').$$

Common weights include $\frac{1}{\sqrt{1-x^2}}$ (Chebyshev–Gauss) and $\sqrt{1-x^2}$. One may also want to integrate over semi-infinite (Gauss-Laguerre quadrature) and infinite intervals (Gauss–Hermite quadrature).

It can be shown that the quadrature nodes x_i are the roots of a polynomial belonging to a class of orthogonal polynomials (the class orthogonal with respect to a weighted inner-product). This is a key observation for computing Gauss quadrature nodes and weights.

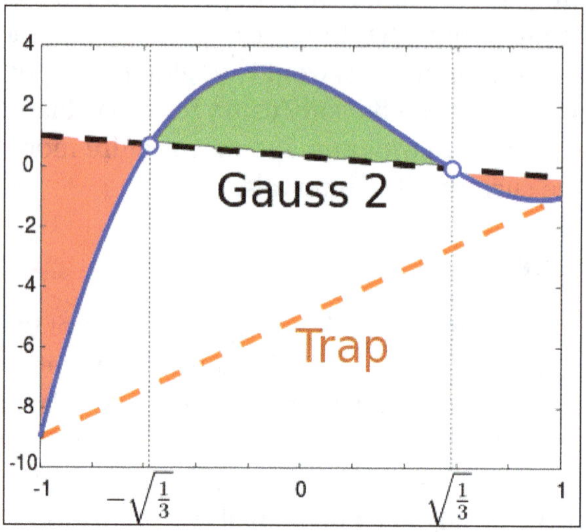

Comparison between 2-point Gaussian and trapezoidal quadrature. The blue line is the polynomial $y(x)=7x^3-8x^2-3x+3$, whose integral in [−1, 1] is $2/3$. The trapezoidal rule returns the integral of the orange dashed line, equal to $y(-1)+y(1)=-10$. The 2-point Gaussian quadrature rule returns the integral of the black dashed curve, equal to $y\left(-\sqrt{\frac{1}{3}}\right)+y\left(\sqrt{\frac{1}{3}}\right)=\frac{2}{3}$. Such a result is exact, since the green region has the same area as the sum of the red regions.

Gauss–Legendre Quadrature

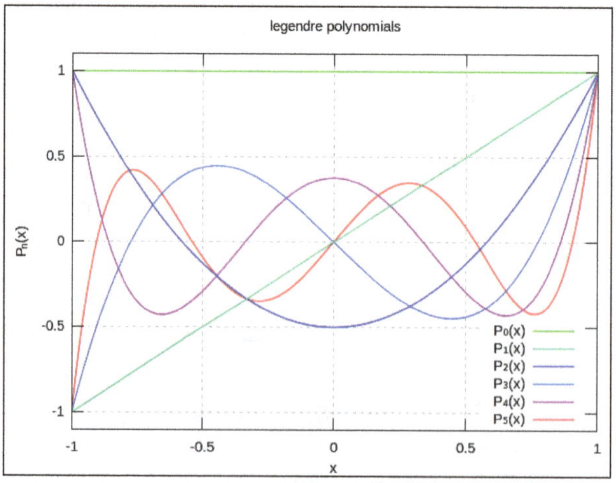

Graphs of Legendre polynomials (up to $n = 5$).

For the simplest integration problem stated above, i.e., $f(x)$ is well-approximated by polynomials on $[-1,1]$, the associated orthogonal polynomials are Legendre polynomials, denoted by $P_n(x)$. With the n-th polynomial normalized to give $P_n(1) = 1$, the i-th Gauss node, x_i, is the i-th root of P_n and the weights are given by the formula.

$$w_i = \frac{2}{\left(1-x_i^2\right)\left[P_n'(x_i)\right]^2}.$$

Some low-order quadrature rules are tabulated below (over interval $[-1, 1]$, see the table for other intervals).

Number of points, n	Points, x_i		Weights, w_i	
1	0		2	
2	$\pm\frac{1}{\sqrt{3}}$	±0.57735...	1	
3	0		$\frac{8}{9}$	0.888889...
	$\pm\sqrt{\frac{3}{5}}$	±0.774597...	$\frac{5}{9}$	0.555556...
4	$\pm\sqrt{\frac{3}{7}-\frac{2}{7}\sqrt{\frac{6}{5}}}$	±0.339981...	$\frac{18+\sqrt{30}}{36}$	0.652145...
	$\pm\sqrt{\frac{3}{7}+\frac{2}{7}\sqrt{\frac{6}{5}}}$	±0.861136...	$\frac{18-\sqrt{30}}{36}$	0.347855...
5	0		$\frac{128}{225}$	0.568889...
	$\pm\frac{1}{3}\sqrt{5-2\sqrt{\frac{10}{7}}}$	±0.538469...	$\frac{322+13\sqrt{70}}{900}$	0.478629...
	$\pm\frac{1}{3}\sqrt{5+2\sqrt{\frac{10}{7}}}$	±0.90618...	$\frac{322-13\sqrt{70}}{900}$	0.236927...

Change of Interval

An integral over $[a, b]$ must be changed into an integral over $[-1, 1]$ before applying the Gaussian quadrature rule. This change of interval can be done in the following way:

$$\int_a^b f(x)dx = \frac{b-a}{2}\int_{-1}^{1} f\left(\frac{b-a}{2}\xi + \frac{a+b}{2}\right)d\xi.$$

Applying n point Gaussian quadrature (ξ, w) rule then results in the following approximation:

$$\int_a^b f(x)dx \approx \frac{b-a}{2}\sum_{i=1}^{n} w_i f\left(\frac{b-a}{2}\xi_i + \frac{a+b}{2}\right).$$

Other Forms

The integration problem can be expressed in a slightly more general way by introducing a positive weight function ω into the integrand, and allowing an interval other than $[-1, 1]$. That is, the problem is to calculate:

$$\int_a^b \omega(x) f(x)dx$$

for some choices of a, b, and ω. For $a = -1$, $b = 1$, and $\omega(x) = 1$, the problem is the same as that considered above. Other choices lead to other integration rules. Some of these are tabulated below. Equation numbers are given for Abramowitz and Stegun (A & S).

Interval	ω(x)	Orthogonal polynomials	A & S
$[-1, 1]$	1	Legendre polynomials	25.4.29
$(-1, 1)$	$(1-x)^{\alpha}(1+x)^{\beta},\quad \alpha,\beta > -1$	Jacobi polynomials	25.4.33 ($\beta = 0$)
$(-1, 1)$	$\frac{1}{\sqrt{1-x^2}}$	Chebyshev polynomials (first kind)	25.4.38
$[-1, 1]$	$\sqrt{1-x^2}$	Chebyshev polynomials (second kind)	25.4.40
$[0, \infty)$	e^{-x}	Laguerre polynomials	25.4.45
$[0, \infty)$	$x^{\alpha}e^{-x},\quad \alpha > -1$	Generalized Laguerre polynomials	
$(-\infty, \infty)$	e^{-x^2}	Hermite polynomials	25.4.46

Fundamental Theorem

Let p_n be a nontrivial polynomial of degree n such that:

$$\int_a^b \omega(x)x^k p_n(x)dx = 0, \quad \textit{for all } k = 0,1,\ldots,n-1.$$

If we pick the n nodes x_i to be the zeros of p_n, then there exist n weights w_i which make the Gauss-quadrature computed integral exact for all polynomials $h(x)$ of degree $2n - 1$ or less. Furthermore, all these nodes x_i will lie in the open interval (a, b).

The polynomial p_n is said to be an orthogonal polynomial of degree n associated to the weight function $\omega(x)$. It is unique up to a constant normalization factor. The idea underlying the proof is that, because of its sufficiently low degree, $h(x)$ can be divided by $p_n(x)$ to produce a quotient $q(x)$ of degree strictly lower than n, and a remainder $r(x)$ of still lower degree, so that both will be orthogonal to $p_n(x)$, by the defining property of $p_n(x)$. Thus:

$$\int_a^b \omega(x)h(x)dx = \int_a^b \omega(x)r(x)dx.$$

Because of the choice of nodes x_i, the corresponding relation:

$$\sum_{i=1}^{n} w_i h(x_i) = \sum_{i=1}^{n} w_i r(x_i)$$

holds also. The exactness of the computed integral for $h(x)$ then follows from corresponding exactness for polynomials of degree only n or less (as is $r(x)$).

General Formula for the Weights

The weights can be expressed as:

$$w_i = \frac{a_n}{a_{n-1}} \frac{\int_a^b \omega(x)p_{n-1}(x)^2\,dx}{p_n'(x_i)p_{n-1}(x_i)}$$

where a_k is the coefficient of x^k in $p_k(x)$. To prove this, note that using Lagrange interpolation one can express $r(x)$ in terms of $r(x_i)$ as:

$$r(x) = \sum_{i=1}^{n} r(x_i) \prod_{\substack{1 \le j \le n \\ j \ne i}} \frac{x - x_j}{x_i - x_j}$$

because $r(x)$ has degree less than n and is thus fixed by the values it attains at n different points. Multiplying both sides by $\omega(x)$ and integrating from a to b yields:

$$\int_a^b \omega(x)r(x)dx = \sum_{i=1}^{n} r(x_i)\int_a^b \omega(x)\prod_{\substack{1\le j\le n\\ j\ne i}} \frac{x-x_j}{x_i-x_j}dx$$

The weights w_i are thus given by:

$$w_i = \int_a^b \omega(x)\prod_{\substack{1\le j\le n\\ j\ne i}} \frac{x-x_j}{x_i-x_j}dx$$

This integral expression for w_i can be expressed in terms of the orthogonal polynomials $p_n(x)$ and $p_{n-1}(x)$ as follows.

We can write:

$$\prod_{\substack{1\le j\le n\\ j\ne i}} \left(x-x_j\right) = \frac{\prod_{1\le j\le n}\left(x-x_j\right)}{x-x_i} = \frac{p_n(x)}{a_n\left(x-x_i\right)}$$

where a_n is the coefficient of x^n in $p_n(x)$. Taking the limit of x to x_i yields using L'Hôpital's rule:

$$\prod_{\substack{1\le j\le n\\ j\ne i}} \left(x-x_j\right) = \frac{p_n'(x_i)}{a_n}$$

We can thus write the integral expression for the weights as:

$$w_i = \frac{1}{p_n'(x_i)}\int_a^b \omega(x)\frac{p_n(x)}{x-x_i}dx$$

In the integrand, writing:

$$\frac{1}{x-x_i} = \frac{1-\left(\frac{x}{x_i}\right)^k}{x-x_i} + \left(\frac{x}{x_i}\right)^k \frac{1}{x-x_i}$$

yields:

$$\int_a^b \omega(x)\frac{x^k p_n(x)}{x-x_i}dx = x_i^k \int_a^b \omega(x)\frac{p_n(x)}{x-x_i}dx$$

provided $k \leq n$, because:

$$\frac{1-\left(\frac{x}{x_i}\right)^k}{x-x_i}$$

is a polynomial of degree k – 1 which is then orthogonal to $p_n(x)$. So, if $q(x)$ is a polynomial of at most nth degree we have:

$$\int_a^b \omega(x)\frac{p_n(x)}{x-x_i}dx = \frac{1}{q(x_i)}\int_a^b \omega(x)\frac{q(x)p_n(x)}{x-x_i}dx$$

We can evaluate the integral on the right hand side for $q(x)=p_{n-1}(x)$ as follows. Because $\frac{p_n(x)}{x-x_i}$ is a polynomial of degree n – 1, we have:

$$\frac{p_n(x)}{x-x_i} = a_n x^{n-1} + s(x)$$

where $s(x)$ is a polynomial of degree $n-2$. Since $s(x)$ is orthogonal to $p_{n-1}(x)$ we have:

$$\int_a^b \omega(x)\frac{p_n(x)}{x-x_i}dx = \frac{a_n}{p_{n-1}(x_i)}\int_a^b \omega(x)p_{n-1}(x)x^{n-1}dx$$

We can then write:

$$x^{n-1} = \left(x^{n-1} - \frac{p_{n-1}(x)}{a_{n-1}}\right) + \frac{p_{n-1}(x)}{a_{n-1}}$$

The term in the brackets is a polynomial of degree $n-2$,, which is therefore orthogonal to $p_{n-1}(x)$. The integral can thus be written as:

$$\int_a^b \omega(x)\frac{p_n(x)}{x-x_i}dx = \frac{a_n}{a_n-1\,p_n-1(x_i)}\int_a^b \omega(x)p_n-1(x)^2\,dx$$

According to equation $w_i = \frac{1}{p_n'(x_i)}\int_a^b \omega(x)\frac{p_n(x)}{x-x_i}dx_i$, the weights are obtained by dividing: this by $p_n'(x_i)$ and that yields the expression in equation $w_i = \frac{a_n}{a_{n-1}}\frac{\int_a^b \omega(x)p_{n-1}(x)^2\,dx}{p_n'(x_i)p_{n-1}(x_i)}$.

w_i can also be expressed in terms of the orthogonal polynomials $p_n(x)$ and now $p_{n+1}(x)$. In the 3-term recurrence relation $p_{n+1}(x_i)=(a)p_n(x_i)+(b)p_{n-1}(x_i)$ the term with $p_n(x_i)$ vanishes, so $p_{n-1}(x_i)$ in equation can be replaced by $\frac{1}{b}p_{n+1}(x_i)$.

Proof that the Weights are Positive

Consider the following polynomial of degree $2n-2$:

$$f(x)=\prod_{\substack{1\le j\le n\\ j\ne i}}\frac{\left(x-x_j\right)^2}{\left(x_i-x_j\right)^2}$$

where, as above, the x_j are the roots of the polynomial $p_n(x)$. Clearly $f(x_j)=\delta_{ij}$. Since the degree of $f(x)$ is less than $2n-1$, the Gaussian quadrature formula involving the weights and nodes obtained from $p_n(x)$ applies. Since $f(x_j)=0$ for j not equal to i, we have:

$$\int_a^b \omega(x)f(x)dx=\sum_{j=1}^{N} w_j f(x_j)=\sum_{j=1}^{N}\delta_{ij}w_j=w_i>0.$$

Since both $\omega(x)$ and $f(x)$ are non-negative functions, it follows that $w_i>0$.

Computation of Gaussian Quadrature Rules

There are many algorithms for computing the nodes x_i and weights w_i of Gaussian quadrature rules. The most popular are the Golub-Welsch algorithm requiring $O(n^2)$ operations, Newton's method for solving $p_n(x)=0$ using the three-term recurrence for evaluation requiring $O(n^2)$ operations, and asymptotic formulas for large n requiring $O(n)$ operations.

Recurrence Relation

Orthogonal polynomials p_r with $(p_r,p_s)=0$ for $r\ne s$ for a scalar product $(,)$, degree $(p_r)=r$ and leading coefficient one (i.e. monic orthogonal polynomials) satisfy the recurrence relation:

$$p_{r+1}(x)=(x-a_{r,r})p_r(x)-a_{r,r-1}p_{r-1}(x)\cdots-a_{r,0}p_0(x)$$

and scalar product defined:

$$(f(x),g(x))=\int_a^b \omega(x)f(x)g(x)dx$$

for $r=0,1,\ldots,n-1$ where n is the maximal degree which can be taken to be infinity, and where $a_{r,s}=\dfrac{(xp_r,p_s)}{(p_s,p_s)}$. First of all, the polynomials defined by the recurrence relation starting with $p_0(x)=1$ have leading coefficient one and correct degree. Given the starting point by p_0, the orthogonality of p_r can be shown by induction. For $r=s=0$ one has:

$$(p_1,p_0)=(x-a_{0,0})(p_0,p_0)=(xp_0,p_0)-a_{0,0}(p_0,p_0)=(xp_0,p_0)-(xp_0,p_0)=0.$$

Now if $p_0, p_1, \dots, p_r$ are orthogonal, then also p_{r+1}, because in:

$$(p_{r+1}, p_s) = (xp_r, p_s) - a_{r,r}(p_r, p_s) - a_{r,r-1}(p_{r-1}, p_s) \cdots - a_{r,0}(p_0, p_s)$$

all scalar products vanish except for the first one and the one where p_s meets the same orthogonal polynomial. Therefore,

$$(p_{r+1}, p_s) = (xp_r, p_s) - a_{r,s}(p_s, p_s) = (xp_r, p_s) - (xp_r, p_s) = 0.$$

However, if the scalar product satisfies $(xf, g) = (f, xg)$ (which is the case for Gaussian quadrature), the recurrence relation reduces to a three-term recurrence relation: For $s < r-1, xp_s i$ is a polynomial of degree less than or equal to $r - 1$. On the other hand, p_r is orthogonal to every polynomial of degree less than or equal to $r - 1$. Therefore, one has $(xp_r, p_s) = (p_r, xp_s) = 0$ and $a_{r,s} = 0$ for $s < r - 1$. The recurrence relation then simplifies to:

$$p_{r+1}(x) = (x - a_{r,r})p_r(x) - a_{r,r-1}p_{r-1}(x)$$

or,

$$p_{r+1}(x) = (x - a_r)p_r(x) - b_r p_{r-1}(x)$$

(with the convention $p_{-1}(x) \equiv 0$) where:

$$a_r := \frac{(xp_r, p_r)}{(p_r, p_r)}, \qquad b_r := \frac{(xp_r, p_{r-1})}{(p_{r-1}, p_{r-1})} = \frac{(p_r, p_r)}{(p_{r-1}, p_{r-1})}$$

(the last because of $(xp_r, p_{r-1}) = (p_r, xp_{r-1}) = (p_r, p_r)$, since xp_{r-1} differs from p_r by a degree less than r).

Golub-Welsch Algorithm

The three-term recurrence relation can be written in matrix form $J\tilde{P} = x\tilde{P} - p_n(x) \times \mathrm{e}_n$ where $\tilde{P} = [p_0(x) \quad p_1(x) \quad \dots \quad p_{n-1}(x)]^\mathrm{T}$, e_n is the nth standard basis vector, i.e. $\mathrm{e}_n = [0 \quad \dots \quad 0 \quad 1]^\mathrm{T}$, and J is the so-called Jacobi matrix:

$$\mathrm{J} = \begin{pmatrix} a_0 & 1 & 0 & \dots & \dots & \dots \\ b_1 & a_1 & 1 & 0 & \dots & \dots \\ 0 & b_2 & a_2 & 1 & 0 & \dots \\ 0 & \dots & \dots & \dots & \dots & 0 \\ \dots & \dots & 0 & b_{n-2} & a_{n-2} & 1 \\ \dots & \dots & \dots & 0 & b_{n-1} & a_{n-1} \end{pmatrix}$$

The zeros x_j of the polynomials up to degree n, which are used as nodes for the Gaussian quadrature can be found by computing the eigenvalues of this tridiagonal matrix. This procedure is known as *Golub–Welsch algorithm*.

For computing the weights and nodes, it is preferable to consider the symmetric tridiagonal matrix $\mathcal{J}$ with elements:

$$\begin{aligned} \mathcal{J}_{i,i} = J_{i,i} &= a_{i-1} & i = 1,\ldots,n \\ \mathcal{J}_{i-1,i} = \mathcal{J}_{i,i-1} = \sqrt{J_{i,i-1}J_{i-1,i}} &= \sqrt{b_{i-1}} & i = 2,\ldots,n. \end{aligned}$$

J and $\mathcal{J}$ are similar matrices and therefore have the same eigenvalues (the nodes). The weights can be computed from the corresponding eigenvectors: If $\phi^{(j)}$ is a normalized eigenvector (i.e., an eigenvector with euclidean norm equal to one) associated to the eigenvalue x_j, the corresponding weight can be computed from the first component of this eigenvector, namely:

$$w_j = \mu_0 (\phi_1^{(j)})^2$$

where μ_0 is the integral of the weight function:

$$\mu_0 = \int_a^b \omega(x)dx.$$

Error Estimates

The error of a Gaussian quadrature rule can be stated as follows. For an integrand which has $2n$ continuous derivatives,

$$\int_a^b \omega(x)f(x)dx - \sum_{i=1}^{n} w_i f(x_i) = \frac{f^{(2n)}(\xi)}{(2n)!}(p_n, p_n)$$

for some ξ in (a, b), where p_n is the monic (i.e. the leading coefficient is 1) orthogonal polynomial of degree n and where:

$$(f,g) = \int_a^b \omega(x)f(x)g(x)dx.$$

In the important special case of $\omega(x) = 1$, we have the error estimate:

$$\frac{(b-a)^{2n+1}(n!)^4}{(2n+1)\left[(2n)!\right]^3} f^{(2n)}(\xi), \qquad a < \xi < b.$$

Stoer and Bulirsch remark that this error estimate is inconvenient in practice, since it may be difficult to estimate the order $2n$ derivative, and furthermore the actual error may be much less than a bound established by the derivative. Another approach is to use two Gaussian quadrature rules of different orders, and to estimate the error as the difference between the two results. For this purpose, Gauss–Kronrod quadrature rules can be useful.

Gauss–Kronrod Rules

If the interval $[a, b]$ is subdivided, the Gauss evaluation points of the new subintervals never coincide with the previous evaluation points (except at zero for odd numbers), and thus the integrand must be evaluated at every point. *Gauss–Kronrod rules* are extensions of Gauss quadrature rules generated by adding $n + 1$ points to an n-point rule in such a way that the resulting rule is of order $2n + 1$. This allows for computing higher-order estimates while re-using the function values of a lower-order estimate. The difference between a Gauss quadrature rule and its Kronrod extension is often used as an estimate of the approximation error.

Gauss–Lobatto Rules

Also known as Lobatto quadrature, named after Dutch mathematician Rehuel Lobatto. It is similar to Gaussian quadrature with the following differences:

- The integration points include the end points of the integration interval.
- It is accurate for polynomials up to degree $2n - 3$, where n is the number of integration points.

Lobatto quadrature of function $f(x)$ on interval $[-1, 1]$:

$$\int_{-1}^{1} f(x)dx = \frac{2}{n(n-1)}[f(1)+f(-1)] + \sum_{i=2}^{n-1} w_i f(x_i) + R_n.$$

Abscissas: x_i is the (i-1) st zero of $P'_{n-1}(x)$.

Weights:

$$w_i = \frac{2}{n(n-1)\left[P_{n-1}(x_i)\right]^2}, \; x_i \neq \pm 1.$$

Remainder:

$$R_n = \frac{-n(n-1)^3 2^{2n-1}\left[(n-2)!\right]^4}{(2n-1)\left[(2n-2)!\right]^3} f^{(2n-2)}(\xi), \qquad -1 < \xi < 1.$$

Some of the weights are:

Number of points, n	Points, x_i	Weights, w_i
3	0	$\frac{4}{3}$
	± 1	$\frac{1}{3}$
4	$\pm\sqrt{\frac{1}{5}}$	$\frac{5}{6}$
	± 1	$\frac{1}{6}$
5	0	$\frac{32}{45}$
	$\pm\sqrt{\frac{3}{7}}$	$\frac{49}{90}$
	± 1	$\frac{1}{10}$
6	$\pm\sqrt{\frac{1}{3}-\frac{2\sqrt{7}}{21}}$	$\frac{14+\sqrt{7}}{30}$
	$\pm\sqrt{\frac{1}{3}+\frac{2\sqrt{7}}{21}}$	$\frac{14-\sqrt{7}}{30}$
	± 1	$\frac{1}{15}$
7	0	$\frac{256}{525}$
	$\pm\sqrt{\frac{5}{11}-\frac{2}{11}\sqrt{\frac{5}{3}}}$	$\frac{124+7\sqrt{15}}{350}$
	$\pm\sqrt{\frac{5}{11}+\frac{2}{11}\sqrt{\frac{5}{3}}}$	$\frac{124-7\sqrt{15}}{350}$
	± 1	$\frac{1}{21}$

An adaptive variant of this algorithm with 2 interior nodes is found in GNU Octave and MATLAB as quadl and integrate.

Lebedev Quadrature

In numerical analysis, Lebedev quadrature, named after Vyacheslav Ivanovich Lebedev, is an approximation to the surface integral of a function over a three-dimensional sphere. The grid is constructed so to have octahedral rotation and inversion symmetry. The number and location of the grid points together with a corresponding set of integration weights are determined by enforcing the exact integration of polynomials (or equivalently, spherical harmonics) up to a given order, leading to a sequence of increasingly dense grids analogous to the one-dimensional Gauss-Legendre scheme.

The Lebedev grid is often employed in the numerical evaluation of volume integrals in the spherical coordinate system, where it is combined with a one-dimensional integration scheme for the radial coordinate. Applications of the grid are found in fields such as computational chemistry and neutron transport.

Angular Integrals

The surface integral of a function over the unit sphere,

$$I[f]=\int d\Omega\, f(\Omega)=\int_0^{\pi}\sin(\theta)d\theta\int_0^{2\pi}d\varphi\, f(\theta,\varphi),$$

is approximated in the Lebedev scheme as:

$$\tilde{I}[f]=4\pi\sum_{i=1}^{N} w_i f(\theta_i,\varphi_i),$$

where the particular grid points and grid weights are to be determined. The use of a single sum, rather than two one dimensional schemes from discretizing the θ and φ integrals individually, leads to more efficient procedure: fewer total grid points are required to obtain similar accuracy. A competing factor is the computational speedup available when using the direct product of two one-dimensional grids. Despite this, the Lebedev grid still outperforms product grids. However, the use of two one-dimensional integration better allows for fine tuning of the grids, and simplifies the use of any symmetry of the integrand to remove symmetry equivalent grid points.

Construction

The Lebedev grid points are constructed so as to lie on the surface of the three-dimensional unit sphere and to be invariant under the octahedral rotation group with inversion. For any point on the sphere, there are either five, seven, eleven, twenty-three, or forty-seven equivalent points with respect to the octahedral group, all of which are included in the grid. Further, all points equivalent under the rotational and inversion group share the same weights. The smallest such set of points is constructed from all six permutations of (±1, 0, 0) (collectively denoted as a^1), leading to an integration scheme:

$$\tilde{I}_6[f] = A_1 \sum_{i=1}^{6} f(a_i^1),$$

Distinct classes of grid points			
	Typical element	Constraint	Number of points
a^1	$(1,0,0)$		6
a^2	$\frac{1}{\sqrt{2}}(1,1,0)$		12
a^3	$\frac{1}{\sqrt{3}}(1,1,1)$		8
b^k	(l_k, l_k, m_k)	$2l_k^2 + m_k^2 = 1$	24
c^k	$(p_k, q_k, 0)$	$p_k^2 + q_k^2 = 1$	24
d^k	(r_k, S_k, W_k)	$r_k^2 + S_k^2 + W_k^2 = 1$	48

where the grid weight is A_1. Geometrically these points correspond to the vertices of a regular octahedron when aligned with the Cartesian axes. Two more sets of points, corresponding to the centers and vertices of the octahedron, are all eight uncorrelated permutations of $1/\sqrt{3}(\pm1,\pm1,\pm1)$ (denoted as a^2), and all twelve permutations of $1/\sqrt{2}(\pm1,\pm1,0)$ (denoted as a^3). This selection of grid points gives rise to the scheme:

$$\tilde{I}_{26}[f] = A_1 \sum_{i=1}^{6} f(a_i^1) + A_2 \sum_{i=1}^{12} f(a_i^2) + A_3 \sum_{i=1}^{8} f(a_i^3),$$

where A_1, A_2, and A_3 are the weight functions that still need to be determined. Three further types of points can be employed as shown in the table. Each of these types of classes can contribute more than one set of points to the grid. In complete generality, the Lebedev scheme is:

$$\tilde{I}_N[f] = A_1 \sum_{i=1}^{6} f(a_i^1) + A_2 \sum_{i=1}^{12} f(a_i^2) + A_3 \sum_{i=1}^{8} f(a_i^3)$$
$$+ \sum_{k=1}^{N_1} B_k \sum_{i=1}^{24} f(b_i^k) + \sum_{k=1}^{N_2} C_k \sum_{i=1}^{24} f(c_i^k) + \sum_{k=1}^{N_3} D_k \sum_{i=1}^{48} f(d_i^k),$$

where the total number of points, N, is:

$$N = 26 + 24(N_1 + N_2) + 48N_3.$$

The determination of the grid weights is achieved by enforcing the scheme to integrate exactly all polynomials up to a given order. On the unit sphere, this is equivalent to integrating all spherical harmonics up to the same order. This problem is simplified by a theorem of Sergei Lvovich Sobolev implying that this condition need be imposed only on those polynomials which are invariant under the octahedral rotation group with inversion. Enforcing these conditions leads to a set of nonlinear equations which have been solved and tabulated up to order 131 in the polynomial.

Newton–Cotes Formulas

In numerical analysis, the Newton–Cotes formulas, also called the Newton–Cotes quadrature rules or simply Newton–Cotes rules, are a group of formulas for numerical integration (also called *quadrature*) based on evaluating the integrand at equally spaced points. They are named after Isaac Newton and Roger Cotes.

Newton–Cotes formulas can be useful if the value of the integrand at equally spaced points is given. If it is possible to change the points at which the integrand is evaluated, then other methods such as Gaussian quadrature and Clenshaw–Curtis quadrature are probably more suitable.

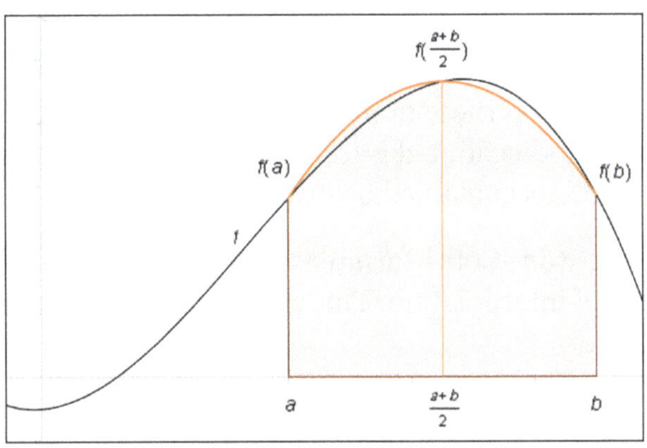

Newton–Cotes formula for $n = 2$.

It is assumed that the value of a function f defined on $[a, b]$ is known at equally spaced points x_i, for $i = 0, \ldots, n$, where $x_0 = a$ and $x_n = b$. There are two types of Newton–Cotes formulas, the "closed" type which uses the function value at all points, and the "open" type which does not use the function values at the endpoints. The closed Newton–Cotes formula of degree n is stated as:

$$\int_a^b f(x)dx \approx \sum_{i=0}^{n} w_i f(x_i)$$

where $x_i = h\,i + x_0$, with h (called the *step size*) equal to $(x_n - x_0)/n = (b-a)/n$. The w_i are called *weights*.

As can be seen in the following derivation the weights are derived from the Lagrange basis polynomials. They depend only on the x_i and not on the function f. Let $L(x)$ be the interpolation polynomial in the Lagrange form for the given data points $(x_0, f(x_0))$, ..., $(x_n, f(x_n))$, then:

$$\int_a^b f(x)dx \approx \int_a^b L(x)dx = \int_a^b \left(\sum_{i=0}^{n} f(x_i) l_i(x) \right) dx = \sum_{i=0}^{n} f(x_i) \underbrace{\int_a^b l_i(x)dx}_{w_i}.$$

The open Newton–Cotes formula of degree n is stated as:

$$\int_a^b f(x)dx \approx \sum_{i=1}^{n-1} w_i f(x_i).$$

The weights are found in a manner similar to the closed formula.

Instability for High Degree

A Newton–Cotes formula of any degree n can be constructed. However, for large n a Newton–Cotes rule can sometimes suffer from catastrophic Runge's phenomenon where the error grows exponentially for large n. Methods such as Gaussian quadrature and Clenshaw–Curtis quadrature with unequally spaced points (clustered at the *endpoints* of the integration interval) are stable and much more accurate, and are normally preferred to Newton–Cotes. If these methods cannot be used, because the integrand is only given at the fixed equidistributed grid, then Runge's phenomenon can be avoided by using a composite rule, as explained below.

Alternatively, stable Newton–Cotes formulas can be constructed using least-squares approximation instead of interpolation. This allows building numerically stable formulas even for high degrees.

Closed Newton–Cotes Formulas

This table lists some of the Newton–Cotes formulas of the closed type. For $0 \le i \le n$, with n the degree, let $x_i = a + i\frac{b-a}{n} = a + ih$, and the notation f_i be a shorthand for $f(x_i)$.

Closed Newton–Cotes Formulas				
Degree n	Step size h	Common name	Formula	Error term
1	$b-a$	Trapezoid rule	$\frac{h}{2}(f_0 + f_1)$	$-\frac{1}{12}h^3 f^{(2)}(\xi)$

2	$\frac{b-a}{2}$	Simpson's rule	$\frac{h}{3}(f_0+4f_1+f_2)$	$-\frac{1}{90}h^5 f^{(4)}(\xi)$
3	$\frac{b-a}{3}$	Simpson's 3/8 rule	$\frac{3h}{8}(f_0+3f_1+3f_2+f_3)$	$-\frac{3}{80}h^5 f^{(4)}(\xi)$
4	$\frac{b-a}{4}$	Boole's rule	$\frac{2h}{45}(7f_0+32f_1+12f_2+32f_3+7f_4)$	$-\frac{8}{945}h^7 f^{(6)}(\xi)$

Boole's rule is sometimes mistakenly called Bode's rule, as a result of the propagation of a typographical error in Abramowitz and Stegun.

The exponent of the segment size $b - a$ in the error term shows the rate at which the approximation error decreases. The degree of the derivative of f in the error term gives the degree up to which polynomials can be integrated exactly (i.e., with error equal to zero) with this rule. Note that the derivative of f in the error term increases by 2 for every other rule. The number ξ must be taken from the interval (a,b).

Open Newton–Cotes formulas

This table lists some of the Newton–Cotes formulas of the open type. Again, f_i is a shorthand for $f(x_i)$, with $x_i = a + i\left(\frac{b-a}{n}\right)$, and n the degree.

Open Newton–Cotes Formulas				
Degree n	Step size h	Common name	Formula	Error term
2	$\frac{b-a}{2}$	Rectangle rule, or midpoint rule	$2hf_1$	$\frac{1}{3}h^3 f^{(2)}(\xi)$
3	$\frac{b-a}{3}$	Trapezoid method	$\frac{3}{2}h(f_1+f_2)$	$\frac{1}{4}h^3 f^{(2)}(\xi)$
4	$\frac{b-a}{4}$	Milne's rule	$\frac{4}{3}h(2f_1-f_2+2f_3)$	$\frac{28}{90}h^5 f^{(4)}(\xi)$
5	$\frac{b-a}{5}$		$\frac{5}{24}h(11f_1+f_2+f_3+11f_4)$	$\frac{95}{144}h^5 f^{(4)}(\xi)$

Composite Rules

For the Newton–Cotes rules to be accurate, the step size h needs to be small, which means that the interval of integration $[a,b]$ must be small itself, which is not true most of the time. For this reason, one usually performs numerical integration by splitting $[a,b]$ into smaller subintervals, applying a Newton–Cotes rule on each subinterval, and adding up the results. This is called a *composite rule.*

Nyström Method

In numerical analysis, the Nyström method or quadrature method seeks the numerical solution of an integral equation by replacing the integral with a representative weighted sum. The continuous problem is broken into n discrete intervals; quadrature or numerical integration determines the weights and locations of representative points for the integral.

The problem becomes a system of linear equations with n equations and n unknowns, and the underlying function is implicitly represented by an interpolation using the chosen quadrature rule. This discrete problem may be ill-conditioned, depending on the original problem and the chosen quadrature rule.

Since the linear equations require $O(n^3)$ operations to solve, high-order quadrature rules perform better because low-order quadrature rules require large n for a given accuracy. Gaussian quadrature is normally a good choice for smooth, non-singular problems.

Discretization of the Integral

Standard quadrature methods seek to represent an integral as a weighed sum in the following manner:

$$\int_a^b h(x)\,dx \approx \sum_{k=1}^{n} w_k h(x_k)$$

where w_k are the weights of the quadrature rule, and points x_k are the abscissas.

Example:

Applying this to the inhomogeneous Fredholm equation of the second kind:

$$f(x) = \lambda u(x) - \int_a^b K(x,x')f(x')\,dx',$$

results in:

$$f(x) \approx \lambda u(x) - \sum_{k=1}^{n} w_k K(x,x_k)f(x_k).$$

Sampling for Nyström Method

The most important step of the Nyström method is sampling, because different sampled landmark points give different approximations of the original matrix.

Uniform sampling without replacement is the most used approach for this purpose, where every point has the same probability of being included in the sample.

Minimum Sum of Squared Similarities Sampling (MSSS)

MSSS is a non-uniform sampling approach that considers both the variance and the similarity of the data distribution in its sampling data, which approximately maximizes the determinant of the reduced similarity matrix that represents the mutual similarities between sampled data points. It is shown in that this approach increase the speed of the clustering on large datasets.

Romberg's Method

In numerical analysis, Romberg's method is used to estimate the definite integral:

$$\int_a^b f(x)dx$$

by applying Richardson extrapolation repeatedly on the trapezium rule or the rectangle rule (midpoint rule). The estimates generate a triangular array. Romberg's method is a Newton–Cotes formula – it evaluates the integrand at equally spaced points. The integrand must have continuous derivatives, though fairly good results may be obtained if only a few derivatives exist. If it is possible to evaluate the integrand at unequally spaced points, then other methods such as Gaussian quadrature and Clenshaw–Curtis quadrature are generally more accurate.

Method

Using:

$$h_n = \tfrac{1}{2^n}(b-a)$$

the method can be inductively defined by:

$$R(0,0) = h_1(f(a)+f(b))$$

$$R(n,0) = \tfrac{1}{2}R(n-1,0)+h_n\sum_{k=1}^{2^{n-1}} f(a+(2k-1)h_n)$$

$$R(n,m) = R(n,m-1)+\tfrac{1}{4^m-1}(R(n,m-1)-R(n-1,m-1))$$

or,

$$R(n,m) = \tfrac{1}{4^m-1}(4^m R(n,m-1)-R(n-1,m-1))$$

where $n \geq m$ and $m \geq 1$. In big O notation, the error for $R(n, m)$ is:

$$O\left(h_n^{2m+2}\right).$$

The zeroeth extrapolation, $R(n, 0)$, is equivalent to the trapezoidal rule with $2^n + 1$ points; the first extrapolation, $R(n, 1)$, is equivalent to Simpson's rule with $2^n + 1$ points. The second extrapolation, $R(n, 2)$, is equivalent to Boole's rule with $2^n + 1$ points. Further extrapolations differ from Newton Cotes formulas. In particular further Romberg extrapolations expand on Boole's rule in very slight ways, modifying weights into ratios similar as in Boole's rule. In contrast, further Newton Cotes methods produce increasingly differing weights, eventually leading to large positive and negative weights. This is indicative of how large degree interpolating polynomial Newton Cotes methods fail to converge for many integrals, while Romberg integration is more stable.

When function evaluations are expensive, it may be preferable to replace the polynomial interpolation of Richardson with the rational interpolation proposed by Bulirsch & Stoer.

Geometric Example

To estimate the area under a curve the trapezoid rule is applied first to one-piece, then two, then four, and so on.

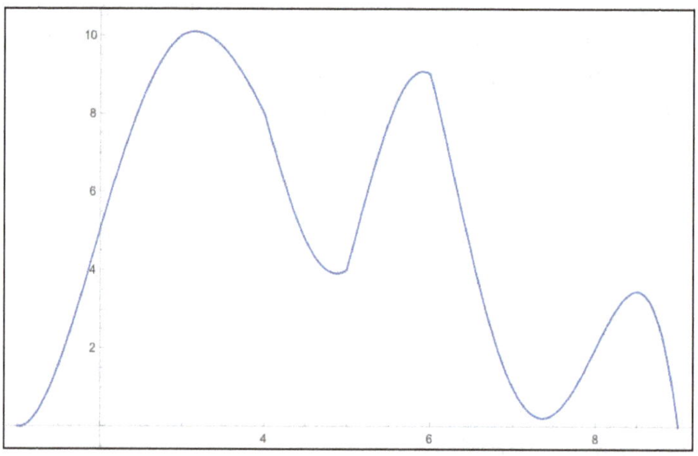

One-piece. Note since it starts and ends at zero, this approximation yields zero area.

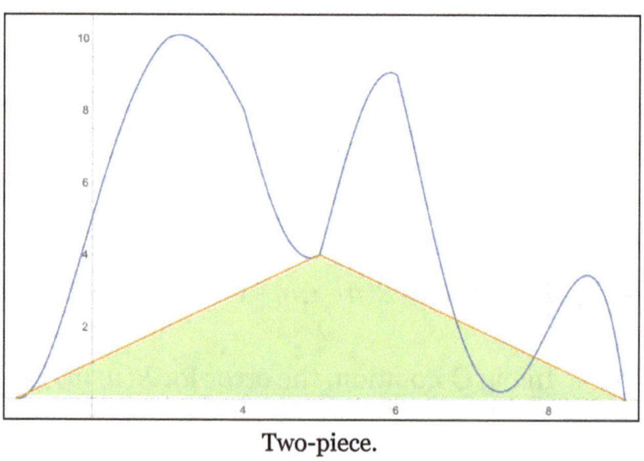

Two-piece.

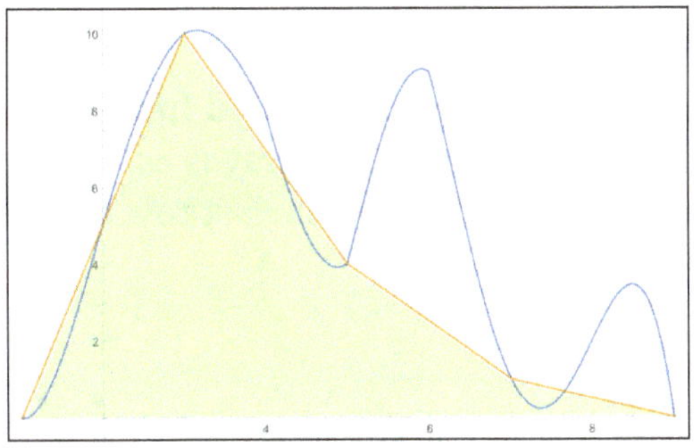

Four-piece.

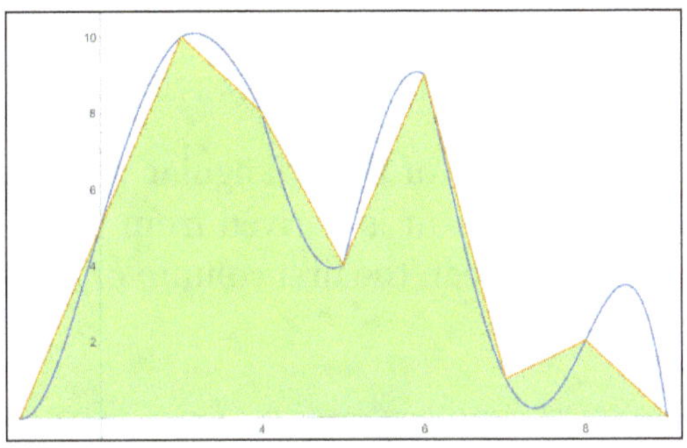

Eight-piece.

After trapezoid rule estimates are obtained, Richardson extrapolation is applied.

- For the first iteration the two piece and one piece estimates are used in the formula (4 × (more accurate) – (less accurate))/3. The same formula is then used to compare the four piece and the two piece estimate, and likewise for the higher estimates.
- For the second iteration the values of the first iteration are used in the formula (16(more accurate) – less accurate))/15.
- The third iteration uses the next power of 4: (64 (more accurate) – less accurate))/63 on the values derived by the second iteration.
- The pattern is continued until there is one estimate.

Number of pieces	Trapezoid estimates	First iteration	Second iteration	Third iteration
		(4 MA – LA)/3*	(16 MA – LA)/15	(64 MA – LA)/63
1	0	(4×16 – 0)/3 = 21.333...	(16×34.667 – 21.333)/15 = 35.556...	(64×42.489 – 35.556)/63 = 42.599...
2	16	(4×30 – 16)/3 = 34.666...	(16×42 – 34.667)/15 = 42.489...	
4	30	(4×39 – 30)/3 = 42		
8	39			

- MA stands for more accurate, LA stands for less accurate.

Example:

As an example, the Gaussian function is integrated from 0 to 1, i.e. the error function erf(1) ≈ 0.842700792949715. The triangular array is calculated row by row and calculation is terminated if the two last entries in the last row differ less than 10^{-8}.

```
0.77174333

0.82526296  0.84310283

0.83836778  0.84273605  0.84271160

0.84161922  0.84270304  0.84270083  0.84270066

0.84243051  0.84270093  0.84270079  0.84270079  0.84270079
```

The result in the lower right corner of the triangular array is accurate to the digits shown. It is remarkable that this result is derived from the less accurate approximations obtained by the trapezium rule in the first column of the triangular array.

Implementation

Here is an example of a computer implementation of the Romberg method (in the C programming language).

```
#include <stdio.h>

#include <math.h>


void

dump_row(size_t i, double *R){

   printf("R[%2zu] = ", i);

   for(size_t j = 0; j <= i; ++j){

      printf("%f ", R[j]);

   }

   printf("\n");

}


double

romberg(double (*f/* function to integrate */)(double), double /*lower
limit*/ a, double /*upper limit*/ b, size_t max_steps, double /*desired
```

```
accuracy*/ acc){
   double R1[max_steps], R2[max_steps]; //buffers
   double *Rp = &R1[0], *Rc = &R2[0]; //Rp is previous row, Rc is cur-
rent row
   double h = (b-a); //step size
   Rp[0] = (f(a) + f(b))*h*.5; //first trapezoidal step

   dump_row(0, Rp);

   for(size_t i = 1; i < max_steps; ++i){
      h /= 2.;
      double c = 0;
      size_t ep = 1 << (i-1); //2^(n-1)
      for(size_t j = 1; j <= ep; ++j){
         c += f(a+(2*j-1)*h);
      }
      Rc[0] = h*c + .5*Rp[0]; //R(i,0)

      for(size_t j = 1; j <= i; ++j){
         double n_k = pow(4, j);
         Rc[j] = (n_k*Rc[j-1] - Rp[j-1])/(n_k-1); //compute R(i,j)
      }

      //Dump ith column of R, R[i,i] is the best estimate so far
      dump_row(i, Rc);

      if(i > 1 && fabs(Rp[i-1]-Rc[i]) < acc){
         return Rc[i-1];
      }
```

```
        //swap Rn and Rc as we only need the last row
        double *rt = Rp;
        Rp = Rc;
        Rc = rt;
    }
    return Rp[max_steps-1]; //return our best guess
}
```

Simpson's Rule

In numerical analysis, Simpson's method is a method for numerical integration, the numerical approximation of definite integrals. Specifically, it is the following approximation for $n+1$ values $x_0 \cdots x_n$ bounding n equally spaced subdivisions (where n is even):

$$\int_a^b f(x)dx \approx \tfrac{\Delta x}{3}\big(f(x_0)+4f(x_1)+2f(x_2)+4f(x_3)+2f(x_4)+\cdots+4f(x_{n-1})+f(x_n)\big),$$

where $\Delta x = \dfrac{b-a}{n}$ and $x_i = a + i\Delta x$.

Simpson's rule also corresponds to the three-point Newton-Cotes quadrature rule.

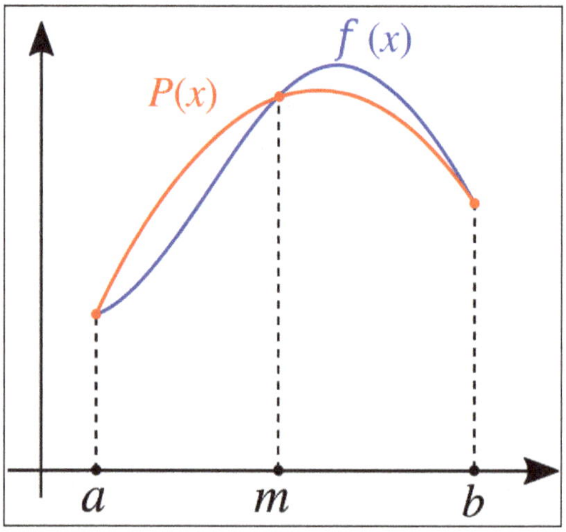

Simpson's rule can be derived by approximating the integrand $f(x)$ (in blue) by the quadratic interpolant $P(x)$ (in red).

Quadratic Interpolation

One derivation replaces the integrand $f(x)$ by the quadratic polynomial (i.e. parabola)

$P(x)$ which takes the same values as $f(x)$ at the end points a and b and the midpoint $m = (a + b) / 2$. One can use Lagrange polynomial interpolation to find an expression for this polynomial,

$$P(x) = f(a)\tfrac{(x-m)(x-b)}{(a-m)(a-b)} + f(m)\tfrac{(x-a)(x-b)}{(m-a)(m-b)} + f(b)\tfrac{(x-a)(x-m)}{(b-a)(b-m)}.$$

Using integration by substitution one can show that:

$$\int_a^b P(x)dx = \tfrac{b-a}{6}\left[f(a) + 4f\left(\tfrac{a+b}{2}\right) + f(b)\right].$$

Introducing the step size $h = (b-a)/2$ this is also commonly written as:

$$\int_a^b p(x)dx = \frac{h}{3}\left[f(a) + 4f\left(\frac{a+b}{2}\right) + f(b)\right].$$

Because of the $1/3$ factor Simpson's rule is also referred to as Simpson's 1/3 rule.

The calculation above can be simplified if one observes that (by scaling) there is no loss of generality in assuming that $a = -1$.

Averaging the Midpoint and the Trapezoidal Rules

Another derivation constructs Simpson's rule from two simpler approximations: the midpoint rule:

$$M = (b-a)f\left(\tfrac{a+b}{2}\right)$$

and the trapezoidal rule:

$$T = \frac{1}{2}(b-a)(f(a) + f(b)).$$

The errors in these approximations are:

$$\frac{1}{24}(b-a)^3 f''(a) + O((b-a)^4) \text{ and } -\frac{1}{12}(b-a)^3 f''(a) + O((b-a)^4),$$

respectively, where $O((b-a)^4)$ denotes a term asymptotically proportional to $(b-a)^4$. The two $O((b-a)^4)$ terms are not equal. It follows from the above formulas for the errors of the midpoint and trapezoidal rule that the leading error term vanishes if we take the weighted average:

$$\frac{2M+T}{3}.$$

This weighted average is exactly Simpson's rule.

Using another approximation (for example, the trapezoidal rule with twice as many points), it is possible to take a suitable weighted average and eliminate another error term. This is Romberg's method.

Undetermined Coefficients

The third derivation starts from the *ansatz:*

$$\tfrac{1}{b-a}\int_a^b f(x)dx \approx \alpha f(a)+\beta f\left(\tfrac{a+b}{2}\right)+\gamma f(b).$$

The coefficients α, β and γ can be fixed by requiring that this approximation be exact for all quadratic polynomials. This yields Simpson's rule.

Error

The error in approximating an integral by Simpson's rule for $n=2$ is:

$$-\tfrac{1}{90}\left(\tfrac{b-a}{2}\right)^5 f^{(4)}(\xi),$$

where ξ is some number between a and b.

The error is asymptotically proportional to $(b-a)^5$.. However, the above derivations suggest an error proportional to $(b-a)^4$. Simpson's rule gains an extra order because the points at which the integrand is evaluated are distributed symmetrically in the interval $[a,b]$.

Since the error term is proportional to the fourth derivative of f at ξ, this shows that Simpson's rule provides exact results for any polynomial f of degree three or less, since the fourth derivative of such a polynomial is zero at all points.

If the second derivative f'' exists and is convex in the interval (a,b):

$$(b-a)f\left(\tfrac{a+b}{2}\right)+\tfrac{1}{3}\left(\tfrac{b-a}{2}\right)^3 f''\left(\tfrac{a+b}{2}\right)\le\int_a^b f(x)dx\le\tfrac{b-a}{6}\left[f(a)+4f\left(\tfrac{a+b}{2}\right)+f(b)\right].$$

Composite Simpson's Rule

If the interval of integration $[a,b]$ is in some sense "small", then Simpson's rule with $n=2$ subintervals will provide an adequate approximation to the exact integral. By small, what we really mean is that the function being integrated is relatively smooth over the interval $[a,b]$. For such a function, a smooth quadratic interpolant like the one used in Simpson's rule will give good results.

However, it is often the case that the function we are trying to integrate is not smooth over the interval. Typically, this means that either the function is highly oscillatory, or it lacks derivatives at certain points. In these cases, Simpson's rule may give very poor results. One common way of handling this problem is by breaking up the interval $[a,b]$ into $n>2$ small subintervals. Simpson's rule is then applied to each subinterval, with the results being summed to produce an approximation for the integral over the entire interval. This sort of approach is termed the *composite Simpson's rule.*

Suppose that the interval $[a,b]$ is split up into n sub-intervals, with n an even number. Then, the composite Simpson's rule is given by:

$$\int_a^b f(x)dx \approx \frac{h}{3}\sum_{j=1}^{n/2}\left[f(x_{2j-2})+4f(x_{2j-1})+f(x_{2j})\right]$$
$$=\frac{h}{3}\left[f(x_0)+2\sum_{j=1}^{n/2-1} f(x_{2j})+4\sum_{j=1}^{n/2} f(x_{2j-1})+f(x_n)\right],$$

where $x_j=a+jh$ for $j=0,1,\ldots,n-1,n$ with $h=(b-a)/n$; in particular, $x_0=a$ and x b. This composite rule with $n=2$ corresponds with the regular Simpson's Rule of the preceding section.

The error committed by the composite Simpson's rule is:

$$-\frac{h^4}{180}(b-a)f^{(4)}(\xi),$$

where ξ is some number between a and b and $h=(b-a)/n$ is the "step length". The error is bounded (in absolute value) by:

$$\tfrac{h^4}{180}(b-a)\max_{\xi\in[a,b]}|f^{(4)}(\xi)|.$$

This formulation splits the interval $[a,b]$ in subintervals of equal length. In practice, it is often advantageous to use subintervals of different lengths, and concentrate the efforts on the places where the integrand is less well-behaved. This leads to the adaptive Simpson's method.

Simpson's 3/8 Rule

Simpson's 3/8 rule is another method for numerical integration proposed by Thomas Simpson. It is based upon a cubic interpolation rather than a quadratic interpolation. Simpson's 3/8 rule is as follows:

$$\int_a^b f(x)dx \approx \frac{3h}{8}\left[f(a)+3f(\frac{2a+b}{3})+3f(\frac{a+2b}{3})+f(b)\right]$$
$$=\frac{(b-a)}{8}\left[f(a)+3f(\frac{2a+b}{3})+3f(\frac{a+2b}{3})+f(b)\right],$$

where $b - a = 3h$. The error of this method is:

$$-\frac{(b-a)^5}{6480} f^{(4)}(\xi),$$

where ξ is some number between a and b. Thus, the 3/8 rule is about twice as accurate as the standard method, but it uses one more function value. A composite 3/8 rule also exists, similarly as above.

A further generalization of this concept for interpolation with arbitrary-degree polynomials are the Newton–Cotes formulas.

Composite Simpson's 3/8 Rule

Dividing the interval $[a,b]$ into n subintervals of length $h = (b-a)/n$ and introducing the nodes $x_i = a + ih$ we have:

$$\begin{aligned}\int_a^b f(x)dx &\approx \frac{3h}{8}[f(x_0)+3f(x_1)+3f(x_2)+2f(x_3)+3f(x_4)+3f(x_5) \\ &+2f(x_6)+\cdots+f(x_n)]. \\ &= \frac{3h}{8}\left[f(x_0)+3\sum_{i\neq 3k}^{n-1} f(x_i)+2\sum_{j=1}^{n/3-1} f(x_{3j})+f(x_n)\right] \qquad \text{For}: k \in \mathbb{N}_0\end{aligned}$$

While the remainder for the rule is shown as:

$$-\frac{h^4}{80}(b-a)f^{(4)}(\xi),$$

We can only use this if n is a multiple of three.

The 3/8th rule is also called Simpson's second rule.

Alternative Extended Simpson's Rule

This is another formulation of a composite Simpson's rule: instead of applying Simpson's rule to disjoint segments of the integral to be approximated, Simpson's rule is applied to overlapping segments, yielding:

$$\int_a^b f(x)dx \approx \tfrac{h}{48}\Big[17f(x_0)+59f(x_1)+43f(x_2)$$

$$+48\sum_{i=4}^{n-4} f(x_i)+49f(x_{n-3})+43f(x_{n-2})+59f(x_{n-1})+17f(x_n)\Big].$$

The formula above is obtained by combining the original composite Simpson's rule with the one consisting of using Simpson's 3/8 rule in the extreme subintervals and the standard 3-point rule in the remaining subintervals. The result is then obtained by taking the mean of the two formulas.

Simpson's Rules in the Case of Narrow Peaks

In the task of estimation of full area of narrow peak-like functions, Simpson's rules are much less efficient than trapezoidal rule. Namely, composite Simpson's 1/3 rule requires 1.8 times more points to achieve the same accuracy as trapezoidal rule. Composite Simpson's 3/8 rule is even less accurate. Integral by Simpson's 1/3 rule can be represented as a sum of 2/3 of integral by trapezoidal rule with step h and 1/3 of integral by rectangle rule with step 2h. No wonder that error of the sum corresponds lo less accurate term. Averaging of Simpson's 1/3 rule composite sums with properly shifted frames produces following rules:

$$\int_a^b f(x)dx \approx \frac{h}{24}[-f(x_{-1})+12f(x_0)+25f(x_1)$$

$$+24\sum_{i=2}^{n-2} f(x_i)+25f(x_{n-1})+12f(x_n)-f(x_{n+1})]$$

where two points outside of integrated region are exploited and,

$$\int_a^b f(x)dx \approx \frac{h}{24}\left[9f(x_0)+28f(x_1)+23f(x_2)\right.$$

$$\left.+24\sum_{i=3}^{n-3} f(x_i)+23f(x_{n-2})+28f(x_{n-1})+9f(x_n)\right]$$

Those rules are very much similar to Press's alternative extended Simpson's rule. Coefficients within the major part of the region being integrated equal one, differences are only at the edges. These three rules can be associated with Euler-MacLaurin formula with the first derivative term and named Euler-MacLaurin integration rules. They differ only in how the first derivative at the region end is calculated.

Composite Simpson's Rule for Irregularly Spaced Data

For some applications, the integration interval $I=[a,b]$ needs to be divided into uneven intervals – perhaps due to uneven sampling of data, or missing or corrupted data points. Suppose we divide the interval I into even number N of subintervals of widths h_k. Then the composite Simpson's rule is given by:

$$\int_a^b f(x)dx = \sum_{i=0}^{N/2-1} \left(\alpha_i f_{2i+2}+\beta_i f_{2i+1}+\eta_i f_{2i}\right),$$

where $f_k = f\left(a + \sum_{i=0}^{k-1} h_i\right)$ are the function values at the kth sampling point on the interval I, and the coefficients α_i, β_i, and η_i are given by"

$$\alpha_i = \frac{2h_{2i+1}^3 - h_{2i}^3 + 3h_{2i}h_{2i+1}^2}{6h_{2i+1}(h_{2i+1} + h_{2i})},$$

$$\beta_i = \frac{h_{2i+1}^3 + h_{2i}^3 + 3h_{2i+1}h_{2i}(h_{2i+1} + h_{2i})}{6h_{2i+1}h_{2i}}, \text{and}$$

$$\eta_i = \frac{2h_{2i}^3 - h_{2i+1}^3 + 3h_{2i+1}h_{2i}^2}{6h_{2i}(h_{2i+1} + h_{2i})}.$$

In case of odd number N of subintervals, the above formula are used up to the second to last interval, and the last interval is handled separately by adding the following to the result:

$$\alpha f_N + \beta f_{N-1} - \eta f_{N-2},$$

where,

$$\alpha = \frac{2h_{N-1}^2 + 3h_{N-1}h_{N-2}}{6(h_{N-2} + h_{N-1})},$$

$$\beta = \frac{h_{N-1}^2 + 3h_{N-1}h_{N-2}}{6h_{N-2}}, and$$

$$\eta = \frac{h_{N-1}^3}{6h_{N-2}(h_{N-2} + h_{N-1})}.$$

Example Implementation in Python

```
import numpy as np


def simpson_nonuniform(x, f) -> float:

    """

    Simpson rule for irregularly spaced data.


        Parameters

        ----------

        x : list or np.array of floats
```

```
            Sampling points for the function values
    f : list or np.array of floats
            Function values at the sampling points

    Returns
    -------
    float : approximation for the integral
"""
N = len(x) - 1
h = np.diff(x)

result = 0.0
for i in range(1, N, 2):
    hph = h[i] + h[i - 1]
    result += f[i] * ( h[i]**3 + h[i - 1]**3
                       + 3. * h[i] * h[i - 1] * hph )\
              / ( 6 * h[i] * h[i - 1] )
    result += f[i - 1] * ( 2. * h[i - 1]**3 - h[i]**3
                          + 3. * h[i] * h[i - 1]**2)\
              / ( 6 * h[i - 1] * hph)
    result += f[i + 1] * ( 2. * h[i]**3 - h[i - 1]**3
                          + 3. * h[i - 1] * h[i]**2)\
              / ( 6 * h[i] * hph )

if (N + 1) % 2 == 0:
    result += f[N] * ( 2 * h[N - 1]**2
                      + 3. * h[N - 2] * h[N - 1])\
              / ( 6 * ( h[N - 2] + h[N - 1] ) )
    result += f[N - 1] * ( h[N - 1]**2
```

```
                    + 3*h[N - 1]* h[N - 2] )\
              / ( 6 * h[N - 2] )
    result -= f[N - 2] * h[N - 1]**3\
              / ( 6 * h[N - 2] * ( h[N - 2] + h[N - 1] ) )
  return result
```

Truncation Error (Numerical Integration)

Truncation error is the difference between a truncated value and the actual value. A truncated quantity is represented by a numeral with a fixed number of allowed digits, with any excess digits "chopped off" (hence the expression "truncated").

As an example of truncation error, consider the speed of light in a vacuum. The official value is 299,792,458 meters per second. In scientific (power-of-10) notation, that quantity is expressed as 2.99792458×10^8. Truncating it to two decimal places yields 2.99×10^8. The truncation error is the difference between the actual value and the truncated value, or 0.00792458×10^8. Expressed properly in scientific notation, it is 7.92458×10^5.

In computing applications, truncation error is the discrepancy that arises from executing a finite number of steps to approximate an infinite process. For example, the infinite series 1/2 + 1/4 + 1/8 + 1/16 + 1/32 ... adds up to exactly 1. However, if we truncate the series to only the first four terms, we get 1/2 + 1/4 + 1/8 + 1/16 = 15/16, producing a truncation error of 1 - 15/16, or 1/16.

References

- NumericalDifferentiation: mathworld.wolfram.com, Retrieved 23 March, 2019
- Types-of-errors-in-numerical-analysis: techwalla.com, Retrieved 02 June, 2019
- Error-mathematics, science: britannica.com, Retrieved 10 January, 2019
- Richard L. Burden, J. Douglas Faires (2000), Numerical Analysis, (7th Ed), Brooks/Cole. ISBN 0-534-38216-9
- Neidinger, Richard D. (2010). "Introduction to Automatic Differentiation and MATLAB Object-Oriented Programming" (PDF). SIAM Review. 52 (3): 545–563. CiteSeerX 10.1.1.362.6580. doi:10.1137/080743627
- Bouneffouf, Djallel; Birol, Inanc (2015), "Sampling with Minimum Sum of Squared Similarities for Nystrom-Based Large Scale Spectral Clustering", Proceedings of the Twenty-Fourth International Joint Conference on Artificial Intelligence, {IJCAI} 2015, Buenos Aires, Argentina, July 25–31, 2015, 8834, AAAI Press, pp. 2313–2319, ISBN 978-1-57735-738-4
- Press, WH; Teukolsky, SA; Vetterling, WT; Flannery, BP (2007), "Section 4.3. Romberg Integration", Numerical Recipes: The Art of Scientific Computing (3rd ed.), New York: Cambridge University Press, ISBN 978-0-521-88068-8
- Truncation-error: whatis.techtarget.com, Retrieved 16 July, 2019

Algorithms in Numerical Analysis

CHAPTER 3

There are various algorithms that are used in numerical analysis such as Brunn's FFT algorithm, Goertzel algorithm, prime factor FFT algorithm, vector-radix FFT algorithm, arbitrary-precision arithmetic, Durand-Kerner method, Brent's method, etc. The topics elaborated in this chapter will help in gaining a better perspective of the different algorithms used in numerical analysis.

Nearly every numerical analysis algorithm has computational complexity that scales exponentially in the underlying physical dimension. Numerical analysis is the subject which studies algorithms for computing expressions defined with real numbers.

Theoretical computer science (TCS) is a subset of computer science that focuses on the mathematical topics as they relate to practical and modern day computing. A widely accepted view of TCS is the study of algorithms and protocols in modern computing. Many theoretical computer courses aim to provide students with a better understanding of fundamental computational techniques.

Components of Computer Science Theory

This field is quite broad and is made up of concepts from an array of other disciplines. Some subjects that compose this theoretical framework include:

- Distributed computation.
- Algorithms.
- Cryptography.
- Algebra.
- Computational Number Theory.
- Data Structures.
- Quantum Computation.
- Machine Learning.
- Computational Biology.

This list is hardly comprehensive. There are many more topics that intersect to make up this diverse field of study. Essentially, though, the theoretical side of computer science and the applied aspects all come together with regard to computing.

It was in 1931 that the mathematician Kurt Godel developed what is known as the incompleteness theorem which concludes that there are definite limitations regarding the kinds of information that be definitively proved or disproved. It is this theory that is the framework for the study of logic and of computability, leading ultimately to the over arching field of computer science.

Other concepts were developed through the years such as information theory, neural networks, parallel distributed frameworks, computational complexity theory and even quantum physics which combined to bring about the modern understanding of theoretical computer science as it's known today.

FFT ALGORITHM

Discrete Fourier Transform

Let there be a polynomial of degree $n-1$:

$$A(x)=a_0x^0+a_1x^1+\cdots+a_{n-1}x^{n-1}$$

Without loss of generality we assume that n-the number of coefficients - is a power of 2. If n s not a power of 2, then we simply add the missing terms a_ix^i and set the coefficients a_i to 0.

The theory of complex numbers tells us that the equation $x^n=1$ has n complex solutions (called the n-th roots of unity), and the solutions are of the form $w_{n,k}=e^{\frac{2k\pi i}{n}}$ with $k=0\ldots n-1$. Additionally these complex numbers have some very interesting properties: e.g. the principal n-th root $w_n=w_{n,1}=e^{\frac{2\pi i}{n}}$ can be used to describe all other n-th roots: $w_{n,k}=(w_n)^k$.

The discrete Fourier transform (DFT) of the polynomial $A(x)$ (or equivalently the vector of coefficients $(a_0,a_1,\ldots,a_{n-1})$ is defined as the values of the polynomial at the points $x=w_{n,k}$ i.e. it is the vector:

$$\begin{aligned}\text{DFT}(a_0,a_1,\ldots,a_{n-1})&=(y_0,y_1,\ldots,y_{n-1})\\&=(A(w_{n,0}),A(w_{n,1}),\ldots,A(w_{n,n-1}))\\&=(A(w_n^0),A(w_n^1),\ldots,A(w_n^{n-1}))\end{aligned}$$

Similarly the inverse discrete Fourier transform is defined: The inverse DFT of values of the polynomial $(y_0, y_1, \ldots, y_{n-1})$ are the coefficients of the polynomial $(a_0, a_1, \ldots, a_{n-1})$.

$$\text{InverseDFT}(y_0, y_1, \ldots, y_{n-1}) = (a_0, a_1, \ldots, a_{n-1})$$

Thus, if a direct DFT computes the values of the polynomial at the points at the n-th roots, the inverse DFT can restore the coefficients of the polynomial using those values.

Application of the DFT: Fast Multiplication of Polynomials

Let there be two polynomials A and B. We compute the DFT for each of them:

$$\text{DFT}(A) \text{ and } \text{DFT}(B).$$

What happens if we multiply these polynomials? Obviously at each point the values are simply multiplied, i.e.

$$(A \cdot B)(x) = A(x) \cdot B(x).$$

This means that if we multiply the vectors $\text{DFT}(A)$ and $\text{DFT}(B)$ - by multiplying each element of one vector by the corresponding element of the other vector - then we get nothing other than the DFT of the polynomial $\text{DFT}(A \cdot B)$:

$$DFT(A \cdot B) = DFT(A) \cdot DFT(B)$$

Finally, applying the inverse DFT, we obtain:

$$A \cdot B = \text{InverseDFT}(\text{DFT}(A) \cdot \text{DFT}(B))$$

On the right the product of the two DFTs we mean the pairwise product of the vector elements. This can be computed in $O(n)$ time. If we can compute the DFT and the inverse DFT in $O(n \log n)$, then we can compute the product of the two polynomials (and consequently also two long numbers) with the same time complexity.

It should be noted, that the two polynomials should have the same degree. Otherwise the two result vectors of the DFT have different length. We can accomplish this by adding coefficients with the value 0.

And also, since the result of the product of two polynomials is a polynomial of degree $2(n-1)$, we have to double the degrees of each polynomial (again by padding 0s). From a vector with n values we cannot reconstruct the desired polynomial with $2n-1$ coefficients.

Fast Fourier Transform

The fast Fourier transform is a method that allows computing the DFT in $O(n \log n)$

time. The basic idea of the FFT is to apply divide and conquer. We divide the coefficient vector of the polynomial into two vectors, recursively compute the DFT for each of them, and combine the results to compute the DFT of the complete polynomial.

So let there be a polynomial $A(x)$ with degree $n-1$, where n is a power of 2, and $n>1$:

$$A(x)=a_0x^0+a_1x^1+\cdots+a_{n-1}x^{n-1}$$

We divide it into two smaller polynomials, the one containing only the coefficients of the even positions, and the one containing the coefficients of the odd positions:

$$A_0(x)=a_0x^0+a_2x^1+\cdots+a_{n-2}x^{\frac{n}{2}}-1$$

$$A_1(x)=a_1x^0+a_3x^1+\cdots+a_{n-1}x^{\frac{n}{2}}-1$$

It is easy to see that:

$$A(x)=A_0(x^2)+xA_1(x^2).$$

The polynomials A_0 and A_1 are only half as much coefficients as the polynomial A. If we can compute the $\text{DFT}(A)$ in linear time using $\text{DFT}(A_0)$ and $\text{DFT}(A_1)$, then we get the recurrence $T_{\text{DFT}}(n)=2T_{\text{DFT}}\left(\frac{n}{2}\right)+O(n)$ for the time complexity, which results in $T_{\text{DFT}}(n)=O(n\log n)$ by the master theorem.

Let's learn how we can accomplish that.

Suppose we have computed the vectors $\left(y_k^0\right)_{k=0}^{n/2-1}=\text{DFT}(A_0)$ and $\left(y_k^1\right)_{k=0}^{n/2-1}=\text{DFT}(A_1)$. Let us find a expression for $\left(y_k\right)_{k=0}^{n-1}=\text{DFT}(A)$.

For the first $\frac{n}{2}$ values we can just use the previously noted equation $A(x)=A_0(x^2)+xA_1(x^2)$:

$$y_k=y_k^0+w_n^ky_k^1,\ k=0\ldots\frac{n}{2}-1.$$

However for the second $\frac{n}{2}$ values we need to find a slightly, different expression:

$$y_{k+n/2}=A\left(w_n^{k+n/2}\right)$$

$$
\begin{aligned}
&= A_0\left(w_n^{2k+n}\right) + w_n^{k+n/2} A_1\left(w_n^{2k+n}\right) \\
&= A_0\left(w_n^{2k} w_n^{n}\right) + w_n^{k} w_n^{n/2} A_1\left(w_n^{2k} w_n^{n}\right) \\
&= A_0\left(w_n^{2k}\right) - w_n^{k} A_1\left(w_n^{2k}\right) \\
&= y_k^0 - w_n^k y_k^1
\end{aligned}
$$

Here we used again $A(x) = A_0(x^2) + xA_1(x^2)$ and the two identities $w_n^n = 1$ and $w_n^{n/2} = -1$ Therefore we get the desired formulas for computing the whole vector (y_k):

$$
y_k = y_k^0 + w_n^k y_k^1, \qquad k = 0\ldots\frac{n}{2}-1,
$$

$$
y_{k+n/2} = y_k^0 - w_n^k y_k^1, \qquad k = 0\ldots\frac{n}{2}-1.
$$

(This pattern $a+b$ and $a-b$ is sometimes called a butterfly.)

Thus we learned how to compute the DFT in $O(n\log n)$ time.

Inverse FFT

Let the vector $(y_0, y_1, \ldots y_{n-1})$ - the values of polynomial A of degree $n-1$ in the points $x = w_n^k$ - be given. We want to restore the coefficients $(a_0, a_1, \ldots, a_{n-1})$ of the polynomial. This known problem is called interpolation, and there are general algorithms for solving it. But in this special case (since we know the values of the points at the roots of unity), we can obtains a much simpler algorithm (that is practically the same as the direct FFT).

We can write the DFT, according to its definition, in the matrix form:

$$
\begin{pmatrix}
w_n^0 & w_n^0 & w_n^0 & w_n^0 & \cdots & w_n^0 \\
w_n^0 & w_n^1 & w_n^2 & w_n^3 & \cdots & w_n^{n-1} \\
w_n^0 & w_n^2 & w_n^4 & w_n^6 & \cdots & w_n^{2(n-1)} \\
w_n^0 & w_n^3 & w_n^6 & w_n^9 & \cdots & w_n^{3(n-1)} \\
\vdots & \vdots & \vdots & \vdots & \ddots & \vdots \\
w_n^0 & w_n^{n-1} & w_n^{2(n-1)} & w_n^{3(n-1)} & \cdots & w_n^{(n-1)(n-1)}
\end{pmatrix}
\begin{pmatrix} a_0 \\ a_1 \\ a_2 \\ a_3 \\ \vdots \\ a_{n-1} \end{pmatrix}
=
\begin{pmatrix} y_0 \\ y_1 \\ y_2 \\ y_3 \\ \vdots \\ y_{n-1} \end{pmatrix}
$$

This matrix is called the Vandermonde matrix.

Thus we can compute the vector $(a_0, a_1, \ldots, a_{n-1})$ by multiplying the vector $(y_0, y_1, \ldots y_{n-1})$ from the left with the inverse of the matrix:

$$\begin{pmatrix} a_0 \\ a_1 \\ a_2 \\ a_3 \\ \vdots \\ a_{n-1} \end{pmatrix} = \begin{pmatrix} w_n^0 & w_n^0 & w_n^0 & w_n^0 & \cdots & w_n^0 \\ w_n^0 & w_n^1 & w_n^2 & w_n^3 & \cdots & w_n^{n-1} \\ w_n^0 & w_n^2 & w_n^4 & w_n^6 & \cdots & w_n^{2(n-1)} \\ w_n^0 & w_n^3 & w_n^6 & w_n^9 & \cdots & w_n^{3(n-1)} \\ \vdots & \vdots & \vdots & \vdots & \ddots & \vdots \\ w_n^0 & w_n^{n-1} & w_n^{2(n-1)} & w_n^{3(n-1)} & \cdots & w_n^{(n-1)(n-1)} \end{pmatrix}^{-1} \begin{pmatrix} y_0 \\ y_1 \\ y_2 \\ y_3 \\ \vdots \\ y_{n-1} \end{pmatrix}$$

A quick check can verify that the inverse of the matrix has the following form:

$$\frac{1}{n} \begin{pmatrix} w_n^0 & w_n^0 & w_n^0 & w_n^0 & \cdots & w_n^0 \\ w_n^0 & w_n^{-1} & w_n^{-2} & w_n^{-3} & \cdots & w_n^{-(n-1)} \\ w_n^0 & w_n^{-2} & w_n^{-4} & w_n^{-6} & \cdots & w_n^{-2(n-1)} \\ w_n^0 & w_n^{-3} & w_n^{-6} & w_n^{-9} & \cdots & w_n^{-3(n-1)} \\ \vdots & \vdots & \vdots & \vdots & \ddots & \vdots \\ w_n^0 & w_n^{-(n-1)} & w_n^{-2(n-1)} & w_n^{-3(n-1)} & \cdots & w_n^{-(n-1)(n-1)} \end{pmatrix}$$

Thus we obtain the formula:

$$a_k = \frac{1}{n} \sum_{j=0}^{n-1} y_j w_n^{-kj}$$

Comparing this to the formula for y_k

$$y_k = \sum_{j=0}^{n-1} a_j w_n^{kj},$$

we notice that these problems are almost the same, so the coefficients a_k can be found by the same divide and conquer algorithm, as well as the direct FFT, only instead of w_n^k we have to use w_n^{-k}, and at the end we need to divide the resulting coefficients by n.

Thus the computation of the inverse DFT is almost the same as the calculation of the direct DFT, and it also can be performed in $O(n \log n)$ time.

Implementation

Here we present a simple recursive implementation of the FFT and the inverse FFT, both in one function, since the difference between the forward and the inverse FFT are so minimal. To store the complex numbers we use the complex type in the C++ STL.

```
using cd = complex<double>;
const double PI = acos(-1);

void fft(vector<cd> & a, bool invert) {
    int n = a.size();
    if (n == 1)
        return;

    vector<cd> a0(n / 2), a1(n / 2);
    for (int i = 0; 2 * i < n; i++) {
        a0[i] = a[2*i];
        a1[i] = a[2*i+1];
    }
    fft(a0, invert);
    fft(a1, invert);

    double ang = 2 * PI / n * (invert ? -1 : 1);
    cd w(1), wn(cos(ang), sin(ang));
    for (int i = 0; 2 * i < n; i++) {
        a[i] = a0[i] + w * a1[i];
        a[i + n/2] = a0[i] - w * a1[i];
        if (invert) {
            a[i] /= 2;
            a[i + n/2] /= 2;
        }
        w *= wn;
    }
}
```

The function gets passed a vector of coefficients, and the function will compute the DFT or inverse DFT and store the result again in this vector. The argument invert shows whether the direct or the inverse DFT should be computed. Inside the function we first check if the length of the vector is equal to one, if this is the case then we don't have to do anything. Otherwise we divide the vector a into two vectors $a0$ and $a1$ and compute the DFT for both recursively. Then we initialize the value wn and a variable w, which will contain the current power of wn. Then the values of the resulting DFT are computed using the above formulas.

If the flag invert is set, then we replace wn with wn^{-1}, and each of the values of the result is divided by 2 (since this will be done in each level of the recursion, this will end up dividing the final values by n).

Using this function we can create a function for multiplying two polynomials:

```
vector<int> multiply(vector<int> const& a, vector<int> const& b) {
    vector<cd> fa(a.begin(), a.end()), fb(b.begin(), b.end());
    int n = 1;
    while (n < a.size() + b.size())
        n <<= 1;
    fa.resize(n);
    fb.resize(n);

    fft(fa, false);
    fft(fb, false);
    for (int i = 0; i < n; i++)
        fa[i] *= fb[i];
    fft(fa, true);

    vector<int> result(n);
    for (int i = 0; i < n; i++)
        result[i] = round(fa[i].real());
    return result;
}
```

This function works with polynomials with integer coefficients, however you can also adjust it to work with other types. Since there is some error when working with complex numbers, we need round the resulting coefficients at the end.

Finally the function for multiplying two long numbers practically doesn't differ from the function for multiplying polynomials. The only thing we have to do afterwards, is to normalize the number:

```
int carry = 0;
for (int i = 0; i < n; i++)
    result[i] += carry;
    carry = result[i] / 10;
    result[i] %= 10;
}
```

Since the length of the product of two numbers never exceed the total length of both numbers, the size of the vector is enough to perform all carry operations.

Improved Implementation: In-place Computation

To increase the efficiency we will switch from the recursive implementation to an iterative one. In the above recursive implementation we explicitly separated the vector a into two vectors - the element on the even positions got assigned to one temporary vector, and the elements on odd positions to another. However if we reorder the elements in a certain way, we don't need to create these temporary vectors (i.e. all the calculations can be done "in-place", right in the vector A itself).

Note that at the first recursion level, the elements whose lowest bit of the position was zero got assigned to the vector a_0, and the ones with a one as the lowest bit of the position got assigned to a_1. In the second recursion level the same thing happens, but with the second lowest bit instead, etc. Therefore if we reverse the bits of the position of each coefficient, and sort them by these reversed values, we get the desired order (it is called the bit-reversal permutation).

For example the desired order for $n=8$ form:

$$a=\{[(a_0,a_4),(a_2,a_6)],[(a_1,a_5),(a_3,a_7)]\}$$

Indeed in the first recursion level (surrounded by curly braces), the vector gets divided into two parts $[a_0,a_2,a_4,a_6]$ and $[a_1,a_3,a_5,a_7]$. As we see, in the bit-reversal permutation this corresponds to simply dividing the vector into two halves: the first $\frac{n}{2}$ elements and the last $\frac{n}{2}$ elements. Then there is a recursive call for each halve. Let the resulting DFT for each of them be returned in place of the elements themselves i.e. the first half and the second half of the vector a respectively.

$$a = \{[y_0^0, y_1^0, y_2^0, y_3^0], [y_0^1, y_1^1, y_2^1, y_3^1]\}$$

Now we want to combine the two DFTs into one for the complete vector. The order of the elements is ideal, and we can also perform the union directly in this vector. We can take the elements y_0^0 and y_0^1 and perform the butterfly transform. The place of the resulting two values is the same as the place of the two initial values, so we get:

$$a = \{[y_0^0 + w_n^0 y_0^1, y_1^0, y_2^0, y_3^0], [y_0^0 - w_n^0 y_0^1, y_1^1, y_2^1, y_3^1]\}$$

Similarly we can compute the butterfly transform of y_1^0 and y_1^1 and put the results in their place, and so on. As a result we get:

$$a = \{[y_0^0 + w_n^0 y_0^1, y_1^0 + w_n^1 y_1^1, y_2^0 + w_n^2 y_2^1, y_3^0 + w_n^3 y_3^1],$$

$$[y_0^0 - w_n^0 y_0^1, y_1^0 - w_n^1 y_1^1, y_2^0 - w_n^2 y_2^1, y_3^0 - w_n^3 y_3^1]\}$$

Thus we computed the required DFT from the vector a.

Here we described the process of computing the DFT only at the first recursion level, but the same works obviously also for all other levels. Thus, after applying the bit-reversal permutation, we can compute the DFT in-place, without any additional memory.

This additionally allows us to get rid of the recursion. We just start at the lowest level, i.e. we divide the vector into pairs and apply the butterfly transform to them. This results with the vector a with the work of the last level applied. In the next step we divide the vector into vectors of size 4, and again apply the butterfly transform, which gives us the DFT for each block of size 4. And so on. Finally in the last step we obtained the result of the DFTs of both halves of a, and by applying the butterfly transform we obtain the DFT for the complete vector a.

```
using cd = complex<double>;

const double PI = acos(-1);


int reverse(int num, int lg_n) {

    int res = 0;

    for (int i = 0; i < lg_n; i++) {

        if (num & (1 << i))
```

```
        res |= 1 << (lg_n - 1 - i);
    }
    return res;
}

void fft(vector<cd> & a, bool invert) {
    int n = a.size();
    int lg_n = 0;
    while ((1 << lg_n) < n)
        lg_n++;

    for (int i = 0; i < n; i++) {
        if (i < reverse(i, lg_n))
            swap(a[i], a[reverse(i, lg_n)]);
    }

    for (int len = 2; len <= n; len <<= 1) {
        double ang = 2 * PI / len * (invert ? -1 : 1);
        cd wlen(cos(ang), sin(ang));
        for (int i = 0; i < n; i += len) {
            cd w(1);
            for (int j = 0; j < len / 2; j++) {
                cd u = a[i+j], v = a[i+j+len/2] * w;
                a[i+j] = u + v;
                a[i+j+len/2] = u - v;
                w *= wlen;
            }
        }
    }
```

```
    if (invert) {
        for (cd & x : a)
            x /= n;
    }
}
```

At first we apply the bit-reversal permutation by swapping the each element with the element of the reversed position. Then the $\log n - 1$ states of the algorithm we compute the DFT for each block of the corresponding size len. For all those blocks we have the same root of unity wlen. We iterate all blocks and perform the butterfly transform on each of them.

We can further optimize the reversal of the bits. In the previous implementation we iterated all bits of the index and created the bitwise reversed index. However we can reverse the bits in a different way.

Suppose that j already contains the reverse of i. Then by to go to $i+1$, we have to increment i, and we also have to increment j, but in a "reversed" number system. Adding one in the conventional binary system is equivalent to flip all tailing ones into zeros and flipping the zero right before them into a one. Equivalently in the "reversed" number system, we flip all leading ones, and the also the next zero.

Thus we get the following implementation:

```
using cd = complex<double>;
const double PI = acos(-1);

void fft(vector<cd> & a, bool invert) {
    int n = a.size();

    for (int i = 1, j = 0; i < n; i++) {
        int bit = n >> 1;
        for (; j & bit; bit >>= 1)
            j ^= bit;
        j ^= bit;
```

```
        if (i < j)
            swap(a[i], a[j]);
    }

    for (int len = 2; len <= n; len <<= 1) {
        double ang = 2 * PI / len * (invert ? -1 : 1);
        cd wlen(cos(ang), sin(ang));
        for (int i = 0; i < n; i += len) {
            cd w(1);
            for (int j = 0; j < len / 2; j++) {
                cd u = a[i+j], v = a[i+j+len/2] * w;
                a[i+j] = u + v;
                a[i+j+len/2] = u - v;
                w *= wlen;
            }
        }
    }

    if (invert) {
        for (cd & x : a)
            x /= n;
    }
}
```

Additionally we can precompute the bit-reversal permutation beforehand. This is especially useful when the size n is the same for all calls. But even when we only have three calls (which are necessary for multiplying two polynomials), the effect is noticeable. Also we can precompute all roots of unity and their powers.

Number Theoretic Transform

Now we switch the objective a little bit. We still want to multiply two polynomials in $O(n \log n)$ time, but this time we want to compute the coefficients modulo some prime number p. Of course for this task we can use the normal DFT and apply the modulo operator to the result. However, doing so might lead to rounding errors, especially when dealing with large numbers. The number theoretic transform (NTT) has the advantage, that it only works with integer, and therefore the result are guaranteed to be correct.

The discrete Fourier transform is based on complex numbers, and the n-th roots of unity. To efficiently compute it, we extensively use properties of the roots (e.g. that there is one root that generates all other roots by exponentiation).

But the same properties hold for the n-th roots of unity in modular arithmetic. A n-th root of unity under a primitive field is such a number w_n that satisfies:

$$(w_n)^n = 1 \pmod{p},$$
$$(w_n)^k \neq 1 \pmod{p}, \quad 1 \leq k < n.$$

The other $n-1$ roots can be obtained as powers of the root w_n.

To apply it in the fast Fourier transform algorithm, we need a root to exist for some n, which is a power of 2, and also for all smaller powers. We can notice the following interesting property:

$$\left(w_n^2\right)^m = w_n^n = 1 \pmod{p}, \quad \text{with } m = \frac{n}{2}$$
$$\left(w_n^2\right)^k = w_n^{2k} \neq 1 \pmod{p}, \quad 1 \leq k < m.$$

Thus if w_n is a n-th root of unity, then w_n^2 is a $\frac{n}{2}$-th root of unity. And consequently for all smaller powers of two there exist roots of the required degree, and they can be computed using w_n.

For computing the inverse DFT, we need the inverse w_n^{-1} of w_n. But for a prime modulus the inverse always exists.

Thus all the properties that we need from the complex roots are also available in modular arithmetic, provided that we have a large enough module p for which a n-th root of unity exists.

For example we can take the following values: module $p = 7340033$, $w_{2^{20}} = 5$. If this module is not enough, we need to find a different pair. We can use that fact that for modules of the form $p = c2^k + 1$ (and p is prime), there always exists the 2^k-th root of unity. It can be shown that g^c is such a 2^k-th root of unity, where g is a primitive root of p.

```
const int mod = 7340033;

const int root = 5;
```

```
const int root_1 = 4404020;
const int root_pw = 1 << 20;

void fft(vector<int> & a, bool invert) {
    int n = a.size();

    for (int i = 1, j = 0; i < n; i++) {
        int bit = n >> 1;
        for (; j & bit; bit >>= 1)
            j ^= bit;
        j ^= bit;

        if (i < j)
            swap(a[i], a[j]);
    }

    for (int len = 2; len <= n; len <<= 1) {
        int wlen = invert ? root_1 : root;
        for (int i = len; i < root_pw; i <<= 1)
            wlen = (int)(1LL * wlen * wlen % mod);

        for (int i = 0; i < n; i += len) {
            int w = 1;
            for (int j = 0; j < len / 2; j++) {
               int u = a[i+j], v = (int)(1LL * a[i+j+len/2] * w % mod);
                a[i+j] = u + v < mod ? u + v : u + v - mod;
                a[i+j+len/2] = u - v >= 0 ? u - v : u - v + mod;
                w = (int)(1LL * w * wlen % mod);
            }
```

```
            }
        }

        if (invert) {
            int n_1 = inverse(n, mod);
            for (int & x : a)
                x = (int)(1LL * x * n_1 % mod);
        }
}
```

Here the function `inverse` computes the modular inverse. The constants `mod`, `root`, `root_pw` determine the module and the root, and `root_1` is the inverse of `root` modulo `mod`.

In practice this implementation is slower than the implementation using complex numbers (due to the huge number of modulo operations), but it has some advantages such as less memory usage and no rounding errors.

Multiplication with Arbitrary Modulus

Multiplying two polynomial $A(x)$, $B(x)$, and computing the coefficients modulo some number M. The number theoretic transform only works for certain prime numbers.

One option would be to perform multiple number theoretic transforms with different prime numbers of the form $c2^k+1$, then apply the Chinese Remainder Theorem to compute the final coefficients.

Another options is to distribute the polynomials $A(x)$ and $B(x)$ into two smaller polynomials each:

$$\begin{aligned} A(x) &= A_1(x) + A_2(x) \cdot C \\ B(x) &= B_1(x) + B_2(x) \cdot C \end{aligned}$$

with $C \approx \sqrt{M}$.

Then the product of $A(x)$ and $B(x)$ can then be represented as:

$$A(x) \cdot B(x) = A_1(x) \cdot B_1(x) + (A_1(x) \cdot B_2(x) + A_2(x) \cdot B_1(x)) \cdot C + (A_2(x) \cdot B_2(x)) \cdot C^2$$

The polynomials $A_1(x)$, $A_2(x)$, $B_1(x)$ and $B_2(x)$ contain only coefficients smaller than

$\sqrt{M}$, therefore the coefficients of all the appearing products are smaller than $M \cdot n$, ,which is usually small enough to handle with typical floating point types.

This approach therefore requires computing the products of polynomials with smaller coefficients (by using the normal FFT and inverse FFT), and then the original product can be restored using modular addition and multiplication in $O(n)$ time.

Applications

DFT can be used in a huge variety of other problems, which at the first glance have nothing to do with multiplying polynomials.

All Possible Sums

We are given two arrays $a[]$ and $b[]$. We have to find all possible sums $a[i]+b[j]$, and for each sum count how often it appears.

For example for $a=[1,2,3]$ and $b=[2,4]$ we get: then sum 3 can be obtained in 1 way, the sum 4 also in 1 way, 5 in 2, 6 in 1, 7 in 1.

We construct for the arrays a and b two polynomials A and B. The numbers of the array will act as the exponents in the polynomial ($a[i] \Rightarrow x^{a[i]}$); and the coefficients of this term will by how often the number appears in the array.

Then, by multiplying these two polynomials in $O(n \log n)$ time, we get a polynomial C, where the exponents will tell us which sums can be obtained, and the coefficients tell us how often. To demonstrate this on the example:

$$(1x^1 + 1x^2 + 1x^3)(1x^2 + 1x^4) = 1x^3 + 1x^4 + 2x^5 + 1x^6 + 1x^7$$

All Possible Scalar Products

We are given two arrays $a[]$ and $b[]$ of length n. We have to compute the products of a with every cyclic shift of b.

We generate two new arrays of size $2n$: We reverse a and append n zeros to it. And we just append b to itself. When we multiply these two arrays as polynomials, and look at the coefficient $c[n-1], c[n], c[2n-2]$ of the product c, we get:

$$c[k] = \sum_{i+j=k} a[i]b[j]$$

And since all the elements $a[i]=0$ for $i \geq n$:

$$c[k] = \sum_{i=0}^{n-1} a[i]b[k-i]$$

It is easy to see that this sum is just the scalar product of the vector a with the $(k-(n-1))$-th cyclic left shift of b. Thus these coefficients are the answer to the problem, and we were still able to obtain it in $O(n \log n)$ time. Note here that $c[2n-1]$ also gives us the n-th cyclic shift but that is the same as the 0-th cyclic shift so we don't need to consider that separately into our answer.

Two Stripes

We are given two Boolean stripes (cyclic arrays of values 0 and 1) a and b. We want to find all ways to attach the first stripe to the second one, such that at no position we have a 1 of the first stripe next to a 1 of the second stripe.

The problem doesn't actually differ much from the previous problem. Attaching two stripes just means that we perform a cyclic shift on the second array, and we can attach the two stripes, if scalar product of the two arrays is 0.

String Matching

We are given two strings, a text T and a pattern P, consisting of lowercase letters. We have to compute all the occurrences of the pattern in the text.

We create a polynomial for each string ($T[i]$ and $P[I]$ are numbers between 0 and 25 corresponding to the 26 letters of the alphabet):

$$A(x) = a_0 x^0 + a_1 x^1 + \cdots + a_{n-1} x^{n-1}, \quad n = |T|$$

with,

$$a_i = \cos(\alpha_i) + i\sin(\alpha_i), \quad \alpha_i = \frac{2\pi T[i]}{26}.$$

And,

$$B(x) = b_0 x^0 + b_1 x^1 + \cdots + b_{m-1} x^{m-1}, \quad m = |P|$$

$$b_i = \cos(\beta_i) - i\sin(\beta_i), \quad \beta_i = \frac{2\pi P[m-i-1]}{26}.$$

Notice that with the expression $P[m-i-1]$ explicitly reverses the pattern.

The $(m-1+i)$ th coefficients of the product of the two polynomials $C(x) = A(x) \cdot B(x)$ will tell us, if the pattern appears in the text at position i.

$$c_{m-1+i} = \sum_{j=0}^{m-1} a_{i+j} \cdot b_{m-1-j} = \sum_{j=0}^{m-1} (\cos(\alpha_{i+j}) + i\sin(\alpha_{i+j})) \cdot (\cos(\beta_j) - i\sin(\beta_j))$$

With $\alpha_{i+j} = \frac{2\pi T[i+j]}{26}$ and $\beta_j = \frac{2\pi P[j]}{26}$

If there is a match, than $T[i+j] = P[j]$ and therefore $\alpha_{i+j} = \beta_j$. This gives (using the Pythagorean trigonometric identity):

$$c_m - 1 + i = \sum_{j=0}^{m-1} a_{i+j} = \left(\cos\left(a_{i+j}\right) + i\,\sin\left(a_{i+j}\right)\right) \cdot \left(\cos\left(a_{i+j}\right) - i\sin\left(a_{i+j}\right)\right)$$
$$= \sum_{j=0}^{m-1} \cos\left(a_{i+j}\right)^2 + \sin\left(a_{i+j}\right)^2 = \sum_{j=0}^{m-1} 1 = m$$

If there isn't a match, then at least a character is different, which leads that one of the products $a_{i+1} \cdot b_{m-1-j}$ is not equal to 1, which leads to the coefficient $c_{m-1+i} \neq m$.

Chirp Z-transform

The chirp Z-transform (CZT) is a generalization of the discrete Fourier transform (DFT). While the DFT samples the Z plane at uniformly-spaced points along the unit circle, the chirp Z-transform samples along spiral arcs in the Z-plane, corresponding to straight lines in the S plane. The DFT, real DFT, and zoom DFT can be calculated as special cases of the CZT.

Specifically, the chirp Z transform calculates the Z transform at a finite number of points z_k along a logarithmic spiral contour, defined as:

$$X_k = \sum_{n=0}^{N-1} x(n) z_k^{-n}$$
$$z_k = A \cdot W^{-k}, k = 0, 1, \ldots, M-1$$

where *A* is the complex starting point, *W* is the complex ratio between points, and *M* is the number of points to calculate.

Like the DFT, the chirp Z-transform can be computed in O(*N* log *N*) operations where *n*=max(M, N). An O(*N* log *N*) algorithm for the inverse chirp Z-transform (ICZT) was described in 2003, and in 2019.

Bluestein's Algorithm

Bluestein's algorithm expresses the CZT as a convolution and implements it efficiently using FFT/IFFT.

As the DFT is a special case of the CZT, this allows the efficient calculation of discrete Fourier transform (DFT) of arbitrary sizes, including prime sizes. (The other algorithm for FFTs of prime sizes, Rader's algorithm, also works by rewriting the DFT as a convolution.) It was conceived in 1968 by Leo Bluestein. Bluestein's algorithm can be used to compute more general transforms than the DFT, based on the (unilateral) z-transform.

Recall that the DFT is defined by the formula:

$$X_k = \sum_{n=0}^{N-1} x_n e^{-\frac{2\pi i}{N}nk} \qquad k = 0,\dots,N-1.$$

If we replace the product nk in the exponent by the identity:

$$nk = \frac{-(k-n)^2}{2} + \frac{n^2}{2} + \frac{k^2}{2}$$

we thus obtain:

$$X_k = e^{-\frac{\pi i}{N}k^2} \sum_{n=0}^{N-1} \left(x_n e^{-\frac{\pi i}{N}n^2} \right) e^{\frac{\pi i}{N}(k-n)^2} \qquad k = 0,\dots,N-1.$$

This summation is precisely a convolution of the two sequences a_n and b_n defined by:

$$a_n = x_n e^{-\frac{\pi i}{N}n^2}$$

$$b_n = e^{\frac{\pi i}{N}n^2},$$

with the output of the convolution multiplied by N phase factors b_k^*. That is:

$$X_k = b_k^* \left(\sum_{n=0}^{N-1} a_n b_{k-n} \right) \qquad k = 0,\dots,N-1.$$

This convolution, in turn, can be performed with a pair of FFTs (plus the pre-computed FFT of complex chirp b_n) via the convolution theorem. The key point is that these FFTs are not of the same length N: such a convolution can be computed exactly from FFTs only by zero-padding it to a length greater than or equal to $2N-1$. In particular, one can pad to a power of two or some other highly composite size, for which the FFT can be efficiently performed by e.g. the Cooley–Tukey algorithm in $O(N \log N)$ time. Thus, Bluestein's algorithm provides an $O(N \log N)$ way to compute prime-size DFTs, albeit several times slower than the Cooley–Tukey algorithm for composite sizes.

The use of zero-padding for the convolution in Bluestein's algorithm deserves some additional comment. Suppose we zero-pad to a length $M \geq 2N-1$. This means that a_n is extended to an array A_n of length M, where $A_n = a_n$ for $0 \leq n < N$ and $A_n = 0$ otherwise—the usual meaning of "zero-padding". However, because of the b_{k-n} term in the convolution, both positive and *negative* values of n are required for b_n (noting that

$b_{-n} = b_n$). The periodic boundaries implied by the DFT of the zero-padded array mean that $-n$ is equivalent to $M-n$. Thus, b_n is extended to an array B_n of length M, where $B_0 = b_0$, $B_n = B_{M-n} = b_n$ for $0 < n < N$, and $B_n = 0$ otherwise. A and B are then FFTed, multiplied pointwise, and inverse FFTed to obtain the convolution of a and b, according to the usual convolution theorem.

Let us also be more precise about what type of convolution is required in Bluestein's algorithm for the DFT. If the sequence b_n were periodic in n with period N, then it would be a cyclic convolution of length N, and the zero-padding would be for computational convenience only. However, this is not generally the case:

$$b_{n+N} = e^{\frac{\pi i}{N}(n+N)^2} = b_n \left[e^{\frac{\pi i}{N}(2Nn+N^2)} \right] = (-1)^N b_n.$$

Therefore, for N even the convolution is cyclic, but in this case N is composite and one would normally use a more efficient FFT algorithm such as Cooley–Tukey. For N odd, however, then b_n is antiperiodic and we technically have a negacyclic convolution of length N. Such distinctions disappear when one zero-pads a_n to a length of at least $2N-1$, however. It is perhaps easiest, therefore, to think of it as a subset of the outputs of a simple linear convolution (i.e. no conceptual "extensions" of the data, periodic or otherwise).

Z-transforms

Bluestein's algorithm can also be used to compute a more general transform based on the (unilateral) z-transform. In particular, it can compute any transform of the form:

$$X_k = \sum_{n=0}^{N-1} x_n z^{nk} \qquad k = 0, \ldots, M-1,$$

for an *arbitrary* complex number z and for *differing* numbers N and M of inputs and outputs. Given Bluestein's algorithm, such a transform can be used, for example, to obtain a more finely spaced interpolation of some portion of the spectrum (although the frequency resolution is still limited by the total sampling time, similar to a Zoom FFT), enhance arbitrary poles in transfer-function analyses, etc.

The algorithm was dubbed the *chirp* z-transform algorithm because, for the Fourier-transform case ($|z| = 1$), the sequence b_n from above is a complex sinusoid of linearly increasing frequency, which is called a (linear) chirp in radar systems.

Bruun's FFT Algorithm

Bruun's algorithm is a fast Fourier transform (FFT) algorithm based on an unusual recursive polynomial-factorization approach, proposed for powers of two by G. Bruun in 1978 and generalized to arbitrary even composite sizes by H. Murakami in 1996. Because its operations involve only real coefficients until the last computation stage, it was initially proposed as a way to efficiently compute the discrete Fourier transform (DFT) of real data. Bruun's algorithm has not seen widespread use, however, as approaches based on the ordinary Cooley–Tukey FFT algorithm have been successfully adapted to real data with at least as much efficiency. Furthermore, there is evidence that Bruun's algorithm may be intrinsically less accurate than Cooley–Tukey in the face of finite numerical precision.

Nevertheless, Bruun's algorithm illustrates an alternative algorithmic framework that can express both itself and the Cooley–Tukey algorithm, and thus provides an interesting perspective on FFTs that permits mixtures of the two algorithms and other generalizations.

Polynomial Approach to the DFT

Recall that the DFT is defined by the formula:

$$X_k = \sum_{n=0}^{N-1} x_n e^{-\frac{2\pi i}{N}nk} \qquad k = 0,\ldots,N-1.$$

For convenience, let us denote the N roots of unity by ω_N^n ($n = 0, \ldots, N-1$):

$$\omega_N^n = e^{-\frac{2\pi i}{N}n}$$

and define the polynomial $x(z)$ whose coefficients are x_n:

$$x(z) = \sum_{n=0}^{N-1} x_n z^n.$$

The DFT can then be understood as a *reduction* of this polynomial; that is, X_k is given by:

$$X_k = x(\omega_N^k) = x(z) \quad \mod(z - \omega_N^k)$$

where mod denotes the polynomial remainder operation. The key to fast algorithms like Bruun's or Cooley–Tukey comes from the fact that one can perform this set of N remainder operations in recursive stages.

Recursive Factorizations and FFTs

In order to compute the DFT, we need to evaluate the remainder of $x(z)$ modulo N degree-1 polynomials. Evaluating these remainders one by one is equivalent to the evaluating the usual DFT formula directly, and requires $O(N^2)$ operations. However, one can *combine* these remainders recursively to reduce the cost, using the following trick: if we want to evaluate $x(z)$ modulo two polynomials $U(z)$ and $V(z)$, we can first take the remainder modulo their product $U(z)V(z)$, which reduces the degree of the polynomial $x(z)$ and makes subsequent modulo operations less computationally expensive.

The product of all of the monomials $(z-\omega_N^k)$ for k=0..N-1 is simply (whose roots are clearly the N roots of unity). One then wishes to find a recursive factorization of z^N-1 into polynomials of few terms and smaller and smaller degree. To compute the DFT, one takes $x(z)$ modulo each level of this factorization in turn, recursively, until one arrives at the monomials and the final result. If each level of the factorization splits every polynomial into an O(1) (constant-bounded) number of smaller polynomials, each with an O(1) number of nonzero coefficients, then the modulo operations for that level take O(N) time; since there will be a logarithmic number of levels, the overall complexity is O (N log N).

More explicitly, suppose for example that $z^N-1=F_1(z)F_2(z)F_3(z)$, and that $F_k(z)=F_{k,1}(z)F_{k,2}(z)$, and so on. The corresponding FFT algorithm would consist of first computing $x_k(z) = x(z) \bmod F_k(z)$, then computing $x_{k,j}(z) = x_k(z) \bmod F_{k,j}(z)$, and so on, recursively creating more and more remainder polynomials of smaller and smaller degree until one arrives at the final degree-0 results.

Moreover, as long as the polynomial factors at each stage are relatively prime (which for polynomials means that they have no common roots), one can construct a dual algorithm by reversing the process with the Chinese Remainder Theorem.

Cooley–Tukey as Polynomial Factorization

The standard decimation-in-frequency (DIF) radix-r Cooley–Tukey algorithm corresponds closely to a recursive factorization. For example, radix-2 DIF Cooley–Tukey factors z^N-1 into $F_1=(z^{N/2}-1)$ and $F_2=(z^{N/2}+1)$. These modulo operations reduce the degree of $x(z)$ by 2, which corresponds to dividing the problem size by 2. Instead of recursively factorizing F_2 directly, though, Cooley–Tukey instead first computes $x_2(z\,\omega_N)$, shifting all the roots (by a *twiddle factor*) so that it can apply the recursive factorization of F_1 to both subproblems. That is, Cooley–Tukey ensures that all subproblems are also DFTs, whereas this is not generally true for an arbitrary recursive factorization.

Bruun Factorization

The basic Bruun algorithm for powers of two N=2^n factorizes $z^{2n}-1$ recursively via the rules:

$$z^{2M}-1=(z^M-1)(z^M+1)$$

$$z^{4M}+az^{2M}+1=(z^{2M}+\sqrt{2-a}z^{M}+1)(z^{2M}-\sqrt{2-a}z^{M}+1)$$

where a is a real constant with $|a| \le 2$. If $a=2\cos(\phi)$, $\phi\in(0,\pi)$, then:

$$\sqrt{2+a}=2\cos\frac{\phi}{2} \text{ and } \sqrt{2-a}=2\cos(\frac{\pi}{2}-\frac{\phi}{2}).$$

At stage s, s=0,1,2,n-1, the intermediate state consists of 2^s polynomials:

$p_{s,0},\ldots,p_{s,2^s-1}$ of degree 2^{n-s} - 1 or less, where:

$$p_{s,0}(z)=p(z)\bmod\left(z^{2^{n-s}}-1\right) \qquad \textit{and}$$
$$p_{s,m}(z)=p(z)\bmod\left(z^{2^{n-s}}-2\cos\left(\tfrac{m}{2^s}\pi\right)z^{2^{n-1-s}}+1\right)m=1,2,\ldots,2^s-1$$

By the construction of the factorization of z^{2n}-1, the polynomials $p_{s,m}(z)$ each encode 2^{n-s} values:

$$X_k=p(e^{2\pi i\frac{k}{2^n}})$$

of the Fourier transform, for m=0, the covered indices are k=0, 2^k, $2\cdot2^s$, $3\cdot2^s$,..., $(2^{n-s}-1)\cdot2^s$, for m>0 the covered indices are $k=m$, $2^{s+1}-m$, $2^{s+1}+m$, $2\cdot2^{s+1}-m$, $2\cdot2^{s+1}+m$, ..., 2^n-m.

During the transition to the next stage, the polynomial $p_{s,\ell}(z)$ is reduced to the polynomials $p_{s+1,\ell}(z)$ and $p_{s+1,2^s-\ell}(z)$ via polynomial division. If one wants to keep the polynomials in increasing index order, this pattern requires an implementation with two arrays. An implementation in place produces a predictable, but highly unordered sequence of indices, for example for N=16 the final order of the 8 linear remainders is (0, 4, 2, 6, 1, 7, 3, 5).

At the end of the recursion, for $s=n$-1, there remain 2^{n-1} linear polynomials encoding two Fourier coefficients X_0 and X_2^{n-1} for the first and for the any other kth polynomial the coefficients X_k and X_{2^n-k}.

At each recursive stage, all of the polynomials of the common degree 4M-1 are reduced to two parts of half the degree 2M-1. The divisor of this polynomial remainder computation is a quadratic polynomial z^m, so that all reductions can be reduced to polynomial divisions of cubic by quadratic polynomials. There are $N/2=2^{n-1}$ of these small divisions at each stage, leading to an O ($N\log N$) algorithm for the FFT.

Moreover, since all of these polynomials have purely real coefficients (until the very last stage), they automatically exploit the special case where the inputs x_n are purely real to save roughly a factor of two in computation and storage. One can also take straightforward advantage of the case of real-symmetric data for computing the discrete cosine transform.

Generalization to Arbitrary Radices

The Bruun factorization, and thus the Bruun FFT algorithm, was generalized to handle arbitrary *even* composite lengths, i.e. dividing the polynomial degree by an arbitrary *radix* (factor), as follows. First, we define a set of polynomials $\varphi_{N,\alpha}(z)$ for positive integers N and for α in [0,1) by:

$$\phi_{N,\alpha}(z) = \begin{cases} z^{2N} - 2\cos(2\pi\alpha)z^N + 1 & \text{if } 0 < \alpha < 1 \\ z^{2N} - 1 & \text{if } \alpha = 0 \end{cases}$$

Note that all of the polynomials that appear in the Bruun factorization above can be written in this form. The zeroes of these polynomials are $e^{2\pi i(\pm\alpha+k)/N}$ for $k = 0,1,\ldots,N-1$ in the $\alpha \neq 0$ case, and $e^{2\pi ik/2N}$ for $k = 0,1,\ldots,2N-1$ in the $\acute{a} = 0$ case. Hence these polynomials can be recursively factorized for a factor (radix) r via:

$$\phi_{rM,\alpha}(z) = \begin{cases} \prod_{\ell=0}^{r-1} \phi_{M,(\alpha+\ell)/r} & \text{if } 0 < \alpha \leq 0.5 \\ \prod_{\ell=0}^{r-1} \phi_{M,(1-\alpha+\ell)/r} & \text{if } 0.5 < \alpha < 1 \\ \prod_{\ell=0}^{r-1} \phi_{M,\ell/(2r)} & \text{if } \alpha = 0 \end{cases}$$

Goertzel Algorithm

The Goertzel algorithm is a technique in digital signal processing (DSP) for efficient evaluation of the individual terms of the discrete Fourier transform (DFT). It is useful in certain practical applications, such as recognition of dual-tone multi-frequency signaling (DTMF) tones produced by the push buttons of the keypad of a traditional analog telephone. The algorithm was first described by Gerald Goertzel in 1958.

Like the DFT, the Goertzel algorithm analyses one selectable frequency component from a discrete signal. Unlike direct DFT calculations, the Goertzel algorithm applies a single real-valued coefficient at each iteration, using real-valued arithmetic for real-valued input sequences. For covering a full spectrum, the Goertzel algorithm has a higher order of complexity than fast Fourier transform (FFT) algorithms, but for computing a small number of selected frequency components, it is more numerically efficient. The simple structure of the Goertzel algorithm makes it well suited to small processors and embedded applications.

The Goertzel algorithm can also be used "in reverse" as a sinusoid synthesis function, which requires only 1 multiplication and 1 subtraction per generated sample.

Algorithm

The main calculation in the Goertzel algorithm has the form of a digital filter, and for this reason the algorithm is often called a *Goertzel filter*. The filter operates on an input sequence $x[n]$ in a cascade of two stages with a parameter ω_0, giving the frequency to be analysed, normalised to radians per sample.

The first stage calculates an intermediate sequence, $s[n]$:

$$s[n] = x[n] + 2\cos(\omega_0)s[n-1] - s[n-2].$$

The second stage applies the following filter to $s[n]$, producing output sequence $y[n]$:

$$y[n] = s[n] - e^{-j\omega_0}s[n-1].$$

The first filter stage can be observed to be a second-order IIR filter with a direct-form structure. This particular structure has the property that its internal state variables equal the past output values from that stage. Input values $x[n]$ for $n<0$ are presumed all equal to 0. To establish the initial filter state so that evaluation can begin at sample $x[0]$, the filter states are assigned initial values $s[-2] = s[-1] = 0$. To avoid aliasing hazards, frequency ω_0 is often restricted to the range 0 to π; using a value outside this range is not meaningless, but is equivalent to using an aliased frequency inside this range, since the exponential function is periodic with a period of 2π in ω_0.

The second-stage filter can be observed to be a FIR filter, since its calculations do not use any of its past outputs.

Z-transform methods can be applied to study the properties of the filter cascade. The Z transform of the first filter stage given in equation $s[n] = x[n] + 2\cos(\omega_0)s[n-1] - s[n-2]$. is:

$$\frac{S(z)}{X(z)} = \frac{1}{1-2\cos(\omega_0)z^{-1}+z^{-2}}$$
$$= \frac{1}{(1-e^{+j\omega_0}z^{-1})(1-e^{-j\omega_0}z^{-1})}.$$

The Z transform of the second filter stage given in equation $y[n] = s[n] - e^{-j\omega_0}s[n-1]$ is:

$$\frac{Y(z)}{S(z)} = 1-e^{-j\omega_0}z^{-1}.$$

The combined transfer function of the cascade of the two filter stages is then:

$$\frac{S(z)}{X(z)}\frac{Y(z)}{S(z)}=\frac{Y(z)}{X(z)}=\frac{(1-e^{-j\omega_0}z^{-1})}{(1-e^{+j\omega_0}z^{-1})(1-e^{-j\omega_0}z^{-1})}$$
$$=\frac{1}{1-e^{+j\omega_0}z^{-1}}.$$

This can be transformed back to an equivalent time-domain sequence, and the terms unrolled back to the first input term at index $n=0$:

$$y[n]=x[n]+e^{+j\omega_0}y[n-1]$$
$$=\sum_{k=-\infty}^{n}x[k]e^{+j\omega_0(n-k)}$$
$$=e^{j\omega_0 n}\sum_{k=0}^{n}x[k]e^{-j\omega_0 k} \qquad \text{since } \forall k<0, x[k]=0.$$

Numerical Stability

It can be observed that the poles of the filter's Z transform are located at $e^{+j\omega_0}$ and $e^{-j\omega_0}$, on a circle of unit radius centered on the origin of the complex Z-transform plane. This property indicates that the filter process is marginally stable and vulnerable to numerical-error accumulation when computed using low-precision arithmetic and long input sequences. A numerically stable version was proposed by Christian Reinsch.

DFT Computations

For the important case of computing a DFT term, the following special restrictions are applied.

- The filtering terminates at index $n=N$, where N is the number of terms in the input sequence of the DFT.
- The frequencies chosen for the Goertzel analysis are restricted to the special form:

  $$\omega_0=2\pi\frac{k}{N}.$$

- The index number k indicating the "frequency bin" of the DFT is selected from the set of index numbers:

  $$k\in\{0,1,2,\ldots,N-1\}.$$

Making these substitutions into equation:

$$[n]=x[n]+e^{+j\omega_0}y[n-1]$$

$$= \sum_{k=-\infty}^{n} x[k]e^{+j\omega_0(n-k)}$$

$$= e^{j\omega_0 n}\sum_{k=0}^{n} x[k]e^{-j\omega_0 k} \qquad \text{since } \forall k<0, x[k]=0.$$

and observing that the term $e^{+j2\pi k}=1$, equation:

$$[n]=x[n]+e^{+j\omega_0}y[n-1]$$

$$= \sum_{k=-\infty}^{n} x[k]e^{+j\omega_0(n-k)}$$

$$= e^{j\omega_0 n}\sum_{k=0}^{n} x[k]e^{-j\omega_0 k} \qquad \text{since } \forall k<0, x[k]=0.$$

then takes the following form:

$$y[N]=\sum_{n=0}^{N} x[n]e^{-j2\pi\frac{nk}{N}}.$$

We can observe that the right side of equation $y[N]=\sum_{n=0}^{N} x[n]e^{-j2\pi\frac{nk}{N}}$.is extremely similar to the defining formula for DFT term $X[k]$, the DFT term for index number k, but not exactly the same. The summation shown in equation $y[N]=\sum_{n=0}^{N} x[n]e^{-j2\pi\frac{nk}{N}}$. requires $N+1$ input terms, but only N input terms are available when evaluating a DFT. A simple but inelegant expedient is to extend the input sequence $x[n]$ with one more artificial value $x[N]=0$. We can see from equation $y[N]=\sum_{n=0}^{N} x[n]e^{-j2\pi\frac{nk}{N}}$.that the mathematical effect on the final result is the same as removing term $x[N]$ from the summation, thus delivering the intended DFT value.

However, there is a more elegant approach that avoids the extra filter pass. From equation $s[n]=x[n]+2\cos(\omega_0)s[n-1]-s[n-2]$, we can note that when the extended input term $x[N]=0$ is used in the final step,

$$s[N]=2\cos(\omega_0)s[N-1]-s[N-2].$$

Thus, the algorithm can be completed as follows:

- Terminate the IIR filter after processing input term $x[N-1]$,
- Apply equation $s[N]=2\cos(\omega_0)s[N-1]-s[N-2]$. to construct $s[N]$ from the prior outputs $s[N-1]$ and $s[N-2]$,

- Apply equation $y[n] = s[n] - e^{-j\omega_0} s[n-1]$. with the calculated $s[N]$ value and with $s[N-1]$ produced by the final direct calculation of the filter.

The last two mathematical operations are simplified by combining them algebraically:

$$
\begin{aligned}
y[N] &= s[N] - e^{-j2\pi\frac{k}{N}} s[N-1] \\
&= (2\cos(\omega_0) s[N-1] - s[N-2]) - e^{-j2\pi\frac{k}{N}} s[N-1] \\
&= e^{j2\pi\frac{k}{N}} s[N-1] - s[N-2].
\end{aligned}
$$

Note that stopping the filter updates at term $N-1$ and immediately applying equation $y[n] = s[n] - e^{-j\omega_0} s[n-1]$. rather than equation:

$$
\begin{aligned}
y[N] &= s[N] - e^{-j2\pi\frac{k}{N}} s[N-1] \\
&= (2\cos(\omega_0) s[N-1] - s[N-2]) - e^{-j2\pi\frac{k}{N}} s[N-1] \\
&= e^{j2\pi\frac{k}{N}} s[N-1] - s[N-2].
\end{aligned}
$$

misses the final filter state updates, yielding a result with incorrect phase.

The particular filtering structure chosen for the Goertzel algorithm is the key to its efficient DFT calculations. We can observe that only one output value $y[N]$ is used for calculating the DFT, so calculations for all the other output terms are omitted. Since the FIR filter is not calculated, the IIR stage calculations $s[0], s[1]$, etc. can be discarded immediately after updating the first stage's internal state.

This seems to leave a paradox: to complete the algorithm, the FIR filter stage must be evaluated once using the final two outputs from the IIR filter stage, while for computational efficiency the IIR filter iteration discards its output values. This is where the properties of the direct-form filter structure are applied. The two internal state variables of the IIR filter provide the last two values of the IIR filter output, which are the terms required to evaluate the FIR filter stage.

Applications

Power-spectrum Terms

Examining equation:

$$y[n] = x[n] + e^{+j\omega_0} y[n-1]$$

$$= \sum_{k=-\infty}^{n} x[k]e^{+j\omega_0(n-k)}$$

$$= e^{j\omega_0 n}\sum_{k=0}^{n} x[k]e^{-j\omega_0 k} \qquad \text{since } \forall k<0, x[k]=0.$$

A final IIR filter pass to calculate term $y[N]$ using a supplemental input value $x[N]=0$ applies a complex multiplier of magnitude 1 to the previous term $y[N-1]$. Consequently, $y[N]$ and $y[N-1]$ represent equivalent signal power. It is equally valid to apply equation:

$$\begin{aligned} y[N] &= s[N] - e^{-j2\pi\frac{k}{N}}s[N-1] \\ &= (2\cos(\omega_0)s[N-1] - s[N-2]) - e^{-j2\pi\frac{k}{N}}s[N-1] \\ &= e^{j2\pi\frac{k}{N}}s[N-1] - s[N-2]. \end{aligned}$$

and calculate the signal power from term $y[N]$ or to apply equation $y[n]= s[n]-e^{-j\omega_0}s[n-1]$. and calculate the signal power from term $y[N-1]$. Both cases lead to the following expression for the signal power represented by DFT term $X[k]$:

$$\begin{aligned} X[k]X'[k] &= y[N]y'[N] = y[N-1]y'[N-1] \\ &= s^2[N-1] + s^2[N-2] - 2\cos\left(2\pi\frac{k}{N}\right)s[N-1]s[N-2]. \end{aligned}$$

In the pseudocode below, the variables `sprev` and `sprev2` temporarily store output history from the IIR filter, while `x[n]` is an indexed element of the array `x`, which stores the input.

```
Nterms defined here

Kterm selected here

ω = 2 × π × Kterm / Nterms;

cr := cos(ω)

ci := sin(ω)

coeff := 2 × cr


sprev := 0

sprev2 := 0
```

```
for each index n in range 0 to Nterms-1 do
    s := x[n] + coeff × sprev - sprev2
    sprev2 := sprev
    sprev := s
end

power := sprev2 × sprev2 + sprev × sprev - coeff × sprev × sprev2
```

It is possible to organise the computations so that incoming samples are delivered singly to a software object that maintains the filter state between updates, with the final power result accessed after the other processing is done.

Single DFT Term with Real-valued Arithmetic

The case of real-valued input data arises frequently, especially in embedded systems where the input streams result from direct measurements of physical processes. Comparing to the illustration in the previous section, when the input data are real-valued, the filter internal state variables sperv and sperv2 can be observed also to be real-valued, consequently, no complex arithmetic is required in the first IIR stage. Optimizing for real-valued arithmetic typically is as simple as applying appropriate real-valued data types for the variables.

After the calculations using input term $x[N-1]$, and filter iterations are terminated, equation:

$$\begin{aligned} y[N] &= s[N] - e^{-j2\pi\frac{k}{N}} s[N-1] \\ &= (2\cos(\omega_0)s[N-1] - s[N-2]) - e^{-j2\pi\frac{k}{N}} s[N-1] \\ &= e^{j2\pi\frac{k}{N}} s[N-1] - s[N-2]. \end{aligned}$$

must be applied to evaluate the DFT term. The final calculation uses complex-valued arithmetic, but this can be converted into real-valued arithmetic by separating real and imaginary terms:

$$c_r = \cos\left(2\pi\frac{k}{N}\right),$$
$$c_i = \sin\left(2\pi\frac{k}{N}\right),$$
$$y[N] = c_r s[N-1] - s[N-2] + jc_i s[N-1].$$

Comparing to the power-spectrum application, the only difference are the calculation used to finish:

```
(Same IIR filter calculations as in the signal power implementation)

XKreal = sprev * cr - sprev2;

XKimag = sprev * ci;
```

Phase Detection

This application requires the same evaluation of DFT term $X[k]$, as discussed in the previous section, using a real-valued or complex-valued input stream. Then the signal phase can be evaluated as:

$$\phi = \tan^{-1} \frac{\Im(X[k])}{\Re(X[k])},$$

aking appropriate precautions for singularities, quadrant, and so forth when computing the inverse tangent function.

Complex Signals in Real Arithmetic

Since complex signals decompose linearly into real and imaginary parts, the Goertzel algorithm can be computed in real arithmetic separately over the sequence of real parts, yielding $y_r[n]$, and over the sequence of imaginary parts, yielding $y_i[n]$. After that, the two complex-valued partial results can be recombined:

$$y[n] = y_r[n] + jy_i[n].$$

Computational Complexity

- According to computational complexity theory, computing a set of M DFT terms using M applications of the Goertzel algorithm on a data set with N values with a "cost per operation" of K has complexity $O(KNM)$.

  To compute a single DFT bin $X(f)$ for a complex input sequence of length N, the Goertzel algorithm requires $2N$ multiplications and $4N$ additions/subtractions within the loop, as well as 4 multiplications and 4 final additions/subtractions, for a total of $2N+4$ multiplications and $4N+4$ additions/subtractions. This is repeated for each of the M frequencies.

- In contrast, using an FFT on a data set with N values has complexity $O(KN \log N)$.

  This is harder to apply directly because it depends on the FFT algorithm used, but a typical example is a radix-2 FFT, which requires $2\log_2(N)$ multiplications and $3\log_2(N)$ additions/subtractions per DFT bin, for each of the N bins.

In the complexity order expressions, when the number of calculated terms M is smaller than $\log N$, the advantage of the Goertzel algorithm is clear. But because FFT code is comparatively complex, the "cost per unit of work" factor K is often larger for an FFT, and the practical advantage favours the Goertzel algorithm even for M several times larger than $\log_2(N)$.

As a rule-of-thumb for determining whether a radix-2 FFT or a Goertzel algorithm is more efficient, adjust the number of terms N in the data set upward to the nearest exact power of 2, calling this N_2, and the Goertzel algorithm is likely to be faster if:

$$M \leq \frac{5N_2}{6N}\log_2(N_2)$$

FFT implementations and processing platforms have a significant impact on the relative performance. Some FFT implementations perform internal complex-number calculations to generate coefficients on-the-fly, significantly increasing their "cost K per unit of work." FFT and DFT algorithms can use tables of pre-computed coefficient values for better numerical efficiency, but this requires more accesses to coefficient values buffered in external memory, which can lead to increased cache contention that counters some of the numerical advantage.

Both algorithms gain approximately a factor of 2 efficiency when using real-valued rather than complex-valued input data. However, these gains are natural for the Goertzel algorithm but will not be achieved for the FFT without using certain algorithm variants specialised for transforming real-valued data.

Prime-factor FFT Algorithm

The prime-factor algorithm (PFA), also called the Good–Thomas algorithm, is a fast Fourier transform (FFT) algorithm that re-expresses the discrete Fourier transform (DFT) of a size $N = N_1N_2$ as a two-dimensional $N_1 \times N_2$ DFT, but *only* for the case where N_1 and N_2 are relatively prime. These smaller transforms of size N_1 and N_2 can then be evaluated by applying PFA recursively or by using some other FFT algorithm.

PFA should not be confused with the mixed-radix generalization of the popular Cooley–Tukey algorithm, which also subdivides a DFT of size $N = N_1N_2$ into smaller transforms of size N_1 and N_2. The latter algorithm can use *any* factors (not necessarily relatively prime), but it has the disadvantage that it also requires extra multiplications by roots of unity called twiddle factors, in addition to the smaller transforms. On the other hand, PFA has the disadvantages that it only works for relatively prime factors (e.g. it is useless for power-of-two sizes) and that it requires a more complicated re-indexing of the data based on the Chinese remainder theorem (CRT). Note, however, that PFA can be combined with mixed-radix Cooley–Tukey, with the former factorizing N into relatively prime components and the latter handling repeated factors.

PFA is also closely related to the nested Winograd FFT algorithm, where the latter performs the decomposed N_1 by N_2 transform via more sophisticated two-dimensional convolution techniques. Some older papers therefore also call Winograd's algorithm a PFA FFT.

Algorithm

Recall that the DFT is defined by the formula:

$$X_k = \sum_{n=0}^{N-1} x_n e^{-\frac{2\pi i}{N} nk} \qquad k = 0, \dots, N-1.$$

The PFA involves a re-indexing of the input and output arrays, which when substituted into the DFT formula transforms it into two nested DFTs (a two-dimensional DFT).

Re-indexing

Suppose that $N = N_1 N_2$, where N_1 and N_2 are relatively prime. In this case, we can define a bijective re-indexing of the input n and output k by:

$$n = n_1 N_2 + n_2 N_1 \quad \bmod N,$$
$$k = k_1 N_2^{-1} N_2 + k_2 N_1^{-1} N_1 \quad \bmod N,$$

where N_1^{-1} denotes the modular multiplicative inverse of N_1 modulo N_2 and vice versa for N_2^{-1}; the indices k_a and n_a run from 0, ..., $N_a - 1$ (for $a = 1, 2$). These inverses only exist for relatively prime N_1 and N_2, and that condition is also required for the first mapping to be bijective.

This re-indexing of n is called the *Ruritanian* mapping (also *Good's* mapping), while this re-indexing of k is called the *CRT* mapping. The latter refers to the fact that k is the solution to the Chinese remainder problem $k = k_1 \bmod N_1$ and $k = k_2 \bmod N_2$.

One could instead use the Ruritanian mapping for the output k and the CRT mapping for the input n, or various intermediate choices.

A great deal of research has been devoted to schemes for evaluating this re-indexing efficiently, ideally in-place, while minimizing the number of costly modulo (remainder) operations.

DFT Re-expression

The above re-indexing is then substituted into the formula for the DFT, and in particular into the product nk in the exponent. Because $e^{2\pi i} = 1$, this exponent is evaluated modulo N: any $N_1 N_2 = N$ cross term in the nk product can be set to zero. (Similarly, X_k and x_n

are implicitly periodic in N, so their subscripts are evaluated modulo N.) The remaining terms give:

$$X_{k_1 N_2^{-1} N_2 + k_2 N_1^{-1} N_1} = \sum_{n_1=0}^{N_1-1} \left(\sum_{n_2=0}^{N_2-1} x_{n_1 N_2 + n_2 N_1} e^{-\frac{2\pi i}{N_2} n_2 k_2} \right) e^{-\frac{2\pi i}{N_1} n_1 k_1}.$$

The inner and outer sums are simply DFTs of size N_2 and N_1, respectively.

Here, we have used the fact that $N_1^{-1}N_1$ is unity when evaluated modulo N_2 in the inner sum's exponent, and vice versa for the outer sum's exponent.

Rader's FFT Algorithm

Rader's algorithm named for Charles M. Rader of MIT Lincoln Laboratory, is a fast Fourier transform (FFT) algorithm that computes the discrete Fourier transform (DFT) of prime sizes by re-expressing the DFT as a cyclic convolution (the other algorithm for FFTs of prime sizes, Bluestein's algorithm, also works by rewriting the DFT as a convolution).

Since Rader's algorithm only depends upon the periodicity of the DFT kernel, it is directly applicable to any other transform (of prime order) with a similar property, such as a number-theoretic transform or the discrete Hartley transform.

The algorithm can be modified to gain a factor of two savings for the case of DFTs of real data, using a slightly modified re-indexing/permutation to obtain two half-size cyclic convolutions of real data; an alternative adaptation for DFTs of real data uses the discrete Hartley transform.

Winograd extended Rader's algorithm to include prime-power DFT sizes p^m, and today Rader's algorithm is sometimes described as a special case of Winograd's FFT algorithm, also called the *multiplicative Fourier transform algorithm* which applies to an even larger class of sizes. However, for composite sizes such as prime powers, the Cooley–Tukey FFT algorithm is much simpler and more practical to implement, so Rader's algorithm is typically only used for large-prime base cases of Cooley–Tukey's recursive decomposition of the DFT.

Algorithm

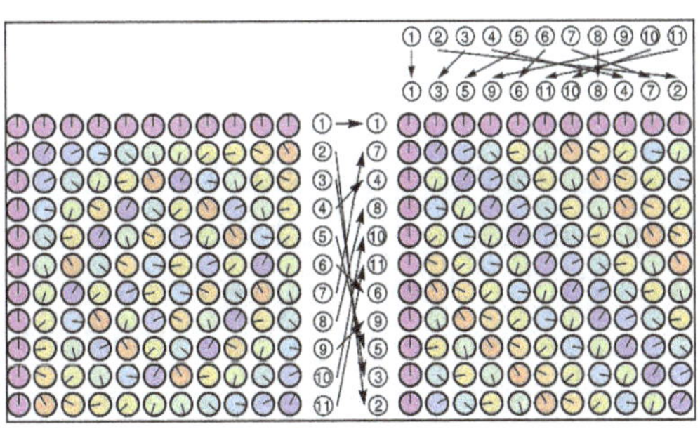

Visual representation of DFT matrix in Rader's FFT algorithm. The array consists of colored clocks represent DFT matrix of size 11. By permuting rows and columns using a sequence generated by the primitive root of 11 (which is 2) except the 1st row and column, the original DFT matrix becomes a circulant matrix. Multiplying a circulant matrix to a data sequence is equivalent to the cyclic convolution. This relation is an example of the fact that multiplicative group is cyclic: $(\mathbb{Z}/p\mathbb{Z})^{\times} \cong C_{p-1}$.

$$X_k = \sum_{n=0}^{N-1} x_n e^{-\frac{2\pi i}{N}nk} \qquad k = 0,\ldots,N-1.$$

If N is a prime number, then the set of non-zero indices $n = 1,\ldots,N-1$ forms a group under multiplication modulo N. One consequence of the number theory of such groups is that there exists a generator of the group (sometimes called a primitive root, which can be found quickly by exhaustive search or slightly better algorithms), an integer g such that $n = g^q \pmod N$ for any non-zero index n and for a unique q in $0,\ldots,N-2$ (forming a bijection from q to non-zero n). Similarly $k = g^{-p} \pmod N$ for any non-zero index k and for a unique p in $0,\ldots,N-2$, where the negative exponent denotes the multiplicative inverse of g^p modulo N. That means that we can rewrite the DFT using these new indices p and q as:

$$X_0 = \sum_{n=0}^{N-1} x_n,$$

$$X_{g^{-p}} = x_0 + \sum_{q=0}^{N-2} x_{g^q} e^{-\frac{2\pi i}{N}g^{-(p-q)}} \qquad p = 0,\ldots,N-2.$$

(Recall that x_n and X_k are implicitly periodic in N, and also that $e^{2\pi i}=1$. Thus, all indices and exponents are taken modulo N as required by the group arithmetic.)

The final summation is precisely a cyclic convolution of the two sequences a_q and b_q of length $N-1$ ($q = 0,\ldots,N-2$) defined by:

$$a_q = x_{g^q}$$

$$b_q = e^{-\frac{2\pi i}{N}g^{-q}}.$$

Evaluating the Convolution

Since $N-1$ is composite, this convolution can be performed directly via the convolution theorem and more conventional FFT algorithms. However, that may not be efficient if $N-1$ itself has large prime factors, requiring recursive use of Rader's algorithm. Instead, one can compute a length-($N-1$) cyclic convolution exactly by zero-padding it to a length of at least $2(N-1)-1$, say to a power of two, which can then be evaluated in $O(N \log N)$ time without the recursive application of Rader's algorithm.

This algorithm, then, requires O(N) additions plus O(N log N) time for the convolution. In practice, the O(N) additions can often be performed by absorbing the additions into the convolution: if the convolution is performed by a pair of FFTs, then the sum of x_n is given by the DC (0th) output of the FFT of a_q plus x_0, and x_0 can be added to all the outputs by adding it to the DC term of the convolution prior to the inverse FFT. Still, this algorithm requires intrinsically more operations than FFTs of nearby composite sizes, and typically takes 3–10 times as long in practice.

If Rader's algorithm is performed by using FFTs of size N–1 to compute the convolution, rather than by zero padding the efficiency depends strongly upon N and the number of times that Rader's algorithm must be applied recursively. The worst case would be if N–1 were 2N_2 where N_2 is prime, with N_2–1 = 2N_3 where N_3 is prime, and so on. In such cases, supposing that the chain of primes extended all the way down to some bounded value, the recursive application of Rader's algorithm would actually require O(N^2) time. Such N_j are called Sophie Germain primes, and such a sequence of them is called a Cunningham chain of the first kind. The lengths of Cunningham chains, however, are observed to grow more slowly than $\log_2(N)$, so Rader's algorithm applied in this way is probably not O(N^2), though it is possibly worse than O(N log N) for the worst cases. Fortunately, a guarantee of O(N log N) complexity can be achieved by zero padding.

Split-radix FFT Algorithm

The split-radix FFT is a fast Fourier transform (FFT) algorithm for computing the discrete Fourier transform (DFT), and was first described in an initially little-appreciated paper by R. Yavne and subsequently rediscovered simultaneously by various authors in 1984. (The name "split radix" was coined by two of these reinventors, P. Duhamel and H. Hollmann.) In particular, split radix is a variant of the Cooley–Tukey FFT algorithm that uses a blend of radices 2 and 4: it recursively expresses a DFT of length N in terms of one smaller DFT of length $N/2$ and two smaller DFTs of length $N/4$.

The split-radix FFT, along with its variations, long had the distinction of achieving the lowest published arithmetic operation count (total exact number of required real additions and multiplications) to compute a DFT of power-of-two sizes N. The arithmetic count of the original split-radix algorithm was improved upon in 2004 (with the initial gains made in unpublished work by J. Van Buskirk via hand optimization for N=64), but it turns out that one can still achieve the new lowest count by a modification of split radix (Johnson and Frigo). Although the number of arithmetic operations is not the sole factor (or even necessarily the dominant factor) in determining the time required to compute a DFT on a computer, the question of the minimum possible count is of longstanding theoretical interest. No tight lower bound on the operation count has currently been proven.

The split-radix algorithm can only be applied when N is a multiple of 4, but since it breaks a DFT into smaller DFTs it can be combined with any other FFT algorithm as desired.

Split-radix Decomposition

Recall that the DFT is defined by the formula:

$$X_k = \sum_{n=0}^{N-1} x_n \omega_N^{nk}$$

where k is an integer ranging from 0 to $N-1$ and ω_N denotes the primitive root of unity:

$$\omega_N = e^{-\frac{2\pi i}{N}},$$

and thus: $\omega_N^N = 1$.

The split-radix algorithm works by expressing this summation in terms of three smaller summations. (Here, we give the "decimation in time" version of the split-radix FFT; the dual decimation in frequency version is essentially just the reverse of these steps.)

First, a summation over the even indices x_{2n_2}. Second, a summation over the odd indices broken into two pieces: x_{4n_4+1} and x_{4n_4+3}, according to whether the index is 1 or 3 modulo 4. Here, n_m denotes an index that runs from 0 to $N/m-1$. The resulting summations look like:

$$X_k = \sum_{n_2=0}^{N/2-1} x_{2n_2} \omega_{N/2}^{n_2 k} + \omega_N^k \sum_{n_4=0}^{N/4-1} x_{4n_4+1} \omega_{N/4}^{n_4 k} + \omega_N^{3k} \sum_{n_4=0}^{N/4-1} x_{4n_4+3} \omega_{N/4}^{n_4 k}$$

where we have used the fact that $\omega_N^{mnk} = \omega_{N/m}^{nk}$. These three sums correspond to *portions* of radix-2 (size N/2) and radix-4 (size N/4) Cooley–Tukey steps, respectively. The underlying idea is that the even-index subtransform of radix-2 has no multiplicative factor in front of it, so it should be left as-is, while the odd-index subtransform of radix-2 benefits by combining a second recursive subdivision.

These smaller summations are now exactly DFTs of length N/2 and N/4, which can be performed recursively and then recombined.

More specifically, let U_k denote the result of the DFT of length N/2 (for $k = 0,\ldots,N/2-1$), and let Z_k and $Z_k^{'}$ denote the results of the DFTs of length N/4 (for $k = 0,\ldots,N/4-1$). Then the output X_k is simply:

$$X_k = U_k + \omega_N^k Z_k + \omega_N^{3k} Z_k^{\prime}.$$

This, however, performs unnecessary calculations, since $k \geq N/4$ turn out to share many calculations with $k < N/4$. In particular, if we add $N/4$ to k, the size-$N/4$ DFTs are not changed (because they are periodic in k), while the size-$N/2$ DFT is unchanged if we add $N/2$ to k. So, the only things that change are the ω_N^k and ω_N^{3k} terms, known as twiddle factors. Here, we use the identities:

$$\omega_N^{k+N/4} = -i\omega_N^k$$
$$\omega_N^{3(k+N/4)} = i\omega_N^{3k}$$

to finally arrive at:

$$X_k = U_k + \left(\omega_N^k Z_k + \omega_N^{3k} Z'_k\right),$$
$$X_{k+N/2} = U_k - \left(\omega_N^k Z_k + \omega_N^{3k} Z'_k\right),$$
$$X_{k+N/4} = U_{k+N/4} - i\left(\omega_N^k Z_k - \omega_N^{3k} Z'_k\right),$$
$$X_{k+3N/4} = U_{k+N/4} + i\left(\omega_N^k Z_k - \omega_N^{3k} Z'_k\right),$$

which gives all of the outputs X_k if we let k range from 0 to $N/4-1$ in the above expressions.

Notice that these expressions are arranged so that we need to combine the various DFT outputs by pairs of additions and subtractions, which are known as butterflies. In order to obtain the minimal operation count for this algorithm, one needs to take into account special cases for $k=0$ (where the twiddle factors are unity) and for $k=N/8$ (where the twiddle factors are $(1\pm i)/\sqrt{2}$ and can be multiplied more quickly). Multiplications by ± 1 and $\pm i$ are ordinarily counted as free (all negations can be absorbed by converting additions into subtractions or vice versa).

This decomposition is performed recursively when N is a power of two. The base cases of the recursion are N=1, where the DFT is just a copy $X_0 = x_0$, and N=2, where the DFT is an addition $X_0 = x_0 + x_1$ and a subtraction $X_1 = x_0 - x_1$.

These considerations result in a count: $4N\log_2 N - 6N + 8$ real additions and multiplications, for N>1 a power of two. This count assumes that, for odd powers of 2, the leftover factor of 2 (after all the split-radix steps, which divide N by 4) is handled directly by the DFT definition (4 real additions and multiplications), or equivalently by a radix-2 Cooley–Tukey FFT step.

Vector-radix FFT Algorithm

The vector-radix FFT algorithm, is a multidimensional fast Fourier transform (FFT) algorithm, which is a generalization of the ordinary Cooley–Tukey FFT algorithm that divides the transform dimensions by arbitrary radices. It breaks a multidimensional

(MD) discrete Fourier transform (DFT) down into successively smaller MD DFTs until, ultimately, only trivial MD DFTs need to be evaluated.

The most common multidimensional FFT algorithm is the row-column algorithm, which means transforming the array first in one index and then in the other. Then a radix-2 direct 2-D FFT has been developed, and it can eliminate 25% of the multiplies as compared to the conventional row-column approach. And this algorithm has been extended to rectangular arrays and arbitrary radices, which is the general vector-radix algorithm.

Vector-radix FFT algorithm can reduce the number of complex multiplications significantly, compared to row-vector algorithm. For example, for a N^M element matrix (M dimensions, and size N on each dimension), the number of complex multiples of vector-radix FFT algorithm for radix-2 is $\frac{2^M-1}{2^M}N^M\log_2 N$, meanwhile, for row-column algorithm, it is $\frac{MN^M}{2}\log_2 N$. And generally, even larger savings in multiplies are obtained when this algorithm is operated on larger radices and on higher dimensional arrays.

Overall, the vector-radix algorithm significantly reduces the structural complexity of the traditional DFT having a better indexing scheme, at the expense of a slight increase in arithmetic operations. So this algorithm is widely used for many applications in engineering, science, and mathematics, for example, implementations in image processing, and high speed FFT processor designing.

2-D DIT Case

As with Cooley–Tukey FFT algorithm, two dimensional vector-radix FFT is derived by decomposing the regular 2-D DFT into sums of smaller DFT's multiplied by "twiddle" factor.

A decimation-in-time (DIT) algorithm means the decomposition is based on time domain x.

We suppose the 2-D DFT:

$$X(k_1,k_2)=\sum_{n_1=0}^{N_1-1}\sum_{n_2=0}^{N_2-1}x[n_1,n_2]\cdot W_{N_1}^{k_1n_1}W_{N_2}^{k_2n_2},$$

where $k_1=0,\ldots,N_1-1$, and $k_2=0,\ldots,N_2-1$, and $x[n_1,n_2]$ is a $N_1\times N_2$ matrix, and $W_N=\exp(-j2\pi/N)$.

For simplicity, let us assume that $N_1=N_2=N$, and radix-$(r\times r)$ (N/r are integers).

Using the change of variables:

- $n_i = rp_i + q_i,$, where $p_i = 0,\ldots,(N/r)-1; q_i = 0,\ldots,r-1;$
- $k_i = u_i + v_i N/r,$, where $u_i = 0,\ldots,(N/r)-1; v_i = 0,\ldots,r-1;$

where $i = 1$ or 2, then the two dimensional DFT can be written as:

$$X(u_1 + v_1 N/r, u_2 + v_2 N/r) = \sum_{q_1=0}^{r-1}\sum_{q_2=0}^{r-1}\left[\sum_{p_1=0}^{N/r-1}\sum_{p_2=0}^{N/r-1} x[rp_1 + q_1, rp_1 + q_1] W_{N/r}^{p_1 u_1} W_{N/r}^{p_2 u_2}\right]\cdot$$

$$W_N^{q_1 u_1 + q_2 u_2} W_r^{q_1 v_1} W_r^{q_2 v_2},$$

The equation above defines the basic structure of the 2-D DIT radix-$(r \times r)$ "butterfly".

When $r = 2,$, the equation can be broken into four summations: one over those samples of x for which both n_1 and n_2 are even, one for which n_1 is even and n_2 is odd, one of which n_1 is odd and n_2 is even, and one for which both n_1 and n_2 are odd, and this leads to:

$$X(k_1,k_2) = S_{00}(k_1,k_2) + S_{01}(k_1,k_2)W_N^{k_2} + S_{10}(k_1,k_2)W_N^{k_1} + S_{11}(k_1,k_2)W_N^{k_1+k_2},$$

where, $S_{ij}(k_1,k_2) = \sum_{n_1=0}^{N/2-1}\sum_{n_2=0}^{N/2-1} x[2n_1 + i, 2n_2 + j]\cdot W_{N/2}^{n_1 k_1} W_{N/2}^{n_2 k_2}.$

2-D DIF Case

Similarly, a decimation-in-frequency (DIF, also called the Sande–Tukey algorithm) algorithm means the decomposition is based on frequency domain X.

Using the change of variables:

- $n_i = p_i + q_i N/r,$ where $p_i = 0,\ldots,(N/r)-1; q_i = 0,\ldots,r-1;$
- $k_i = ru_i + v_i,$ where $u_i = 0,\ldots,(N/r)-1; v_i = 0,\ldots,r-1;$

where $i = 1$ or 2, and the DFT equation can be written as:

$$X(ru_1 + v_1, ru_2 + v_2) = \sum_{p_1=0}^{N/r-1}\sum_{p_2=0}^{N/r-1}\left[\sum_{q_1=0}^{r-1}\sum_{q_2=0}^{r-1} x[p_1 + q_1 N/r, p_1 + q_1 N/r] W_r^{q_1 v_1} W_r^{q_2 v_2}\right]\cdot$$

$$W_N^{p_1 v_1 + p_2 v_2} W_{N/r}^{p_1 u_1} W_{N/r}^{p_2 u_2},$$

Other Approaches

The split-radix FFT algorithm has been proved to be a useful method for 1-D DFT. And this method has been applied to the vector-radix FFT to obtain a split vector-radix FFT.

In conventional 2-D vector-radix algorithm, we decompose the indices k_1, k_2 into 4 groups:

$$\begin{aligned} X(2k_1, 2k_2) &\quad : \quad \textit{even-even} \\ X(2k_1, 2k_2+1) &\quad : \quad \textit{even-odd} \\ X(2k_1+1, 2k_2) &\quad : \quad \textit{odd-even} \\ X(2k_1+1, 2k_2+1) &\quad : \quad \textit{odd-odd} \end{aligned}$$

By the split vector-radix algorithm, the first three groups remain unchanged, the fourth odd-odd group is further decomposed into another four sub-groups, and seven groups in total:

$$\begin{aligned} X(2k_1, 2k_2) &\quad : \quad \textit{even-even} \\ X(2k_1, 2k_2+1) &\quad : \quad \textit{even-odd} \\ X(2k_1+1, 2k_2) &\quad : \quad \textit{odd-even} \\ X(4k_1+1, 4k_2+1) &\quad : \quad \textit{odd-odd} \\ X(4k_1+1, 4k_2+3) &\quad : \quad \textit{odd-odd} \\ X(4k_1+3, 4k_2+1) &\quad : \quad \textit{odd-odd} \\ X(4k_1+3, 4k_2+3) &\quad : \quad \textit{odd-odd} \end{aligned}$$

That means the fourth term in 2-D DIT radix-(2×2) equation, $S_{11}(k_1, k_2)W_N^{k_1+k_2}$ becomes:

$$A_{11}(k_1, k_2)W_N^{k_1+k_2} + A_{13}(k_1, k_2)W_N^{k_1+3k_2} + A_{31}(k_1, k_2)W_N^{3k_1+k_2} + A_{33}(k_1, k_2)W_N^{3(k_1+k_2)},$$

where,

$$A_{ij}(k_1, k_2) = \sum_{n_1=0}^{N/4-1} \sum_{n_2=0}^{N/4-1} x[4n_1+i, 4n_2+j] \cdot W_{N/4}^{n_1 k_1} W_{N/4}^{n_2 k_2}$$

The 2-D N by N DFT is then obtained by successive use of the above decomposition, up to the last stage.

It has been shown that the split vector radix algorithm has saved about 30% of the complex multiplications and about the same number of the complex additions for typical 1024×1024 array, compared with the vector-radix algorithm.

COMPUTER ARITHMETIC ALGORITHMS

Addition-chain Exponentiation

In mathematics and computer science, optimal addition-chain exponentiation is a method of exponentiation by positive integer powers that requires a minimal number of multiplications. This corresponds to the sequence A003313 on the Online Encyclopedia of Integer Sequences. It works by creating the shortest addition chain that generates the desired exponent. Each exponentiation in the chain can be evaluated by multiplying two of the earlier exponentiation results. More generally, *addition-chain exponentiation* may also refer to exponentiation by non-minimal addition chains constructed by a variety of algorithms (since a shortest addition chain is very difficult to find).

The shortest addition-chain algorithm requires no more multiplications than binary exponentiation and usually less. The first example of where it does better is for a^{15}, where the binary method needs six multiplications but a shortest addition chain requires only five:

$$a^{15} = a \times (a \times [a \times a^2]^2)^2 \text{ (binary, 6 multiplications)}$$

$$a^{15} = a^3 \times ([a^3]^2)^2 \text{ (shortest addition chain, 5 multiplications).}$$

Table: Demonstrating how to do exponentiation using addition chains.

Number of Multiplications	Actual Exponentiation	Specific implementation of Addition Chains to do Exponentiation
0	a^1	a
1	a^2	a × a
2	a^3	a × a × a
2	a^4	(a × a→b) × b
3	a^5	(a × a→b) × b × a
3	a^6	(a × a→b) × b × b
4	a^7	(a × a→b) × b × b × a
3	a^8	((a × a→b) × b→d) × d
4	a^9	(a × a × a→c) × c × c
4	a^{10}	((a × a→b) × b→d) × d × b
5	a^{11}	((a × a→b) × b→d) × d × b × a
4	a^{12}	((a × a→b) × b→d) × d × d
5	a^{13}	((a × a→b) × b→d) × d × d × a
5	a^{14}	((a × a→b) × b→d) × d × d × b
5	a^{15}	((a × a→b) × b × a→e) × e × e
4	a^{16}	(((a × a→b) × b→d) × d→h) × h

On the other hand, the determination of a shortest addition chain is hard: no efficient optimal methods are currently known for arbitrary exponents, and the related problem of finding a shortest addition chain for a given set of exponents has been proven NP-complete. Even given a shortest chain, addition-chain exponentiation requires more memory than the binary method, because it must potentially store many previous exponents from the chain. So in practice, shortest addition-chain exponentiation is primarily used for small fixed exponents for which a shortest chain can be precomputed and is not too large.

There are also several methods to *approximate* a shortest addition chain, and which often require fewer multiplications than binary exponentiation; binary exponentiation itself is a suboptimal addition-chain algorithm. The optimal algorithm choice depends on the context (such as the relative cost of the multiplication and the number of times a given exponent is re-used).

The problem of finding the shortest addition chain cannot be solved by dynamic programming, because it does not satisfy the assumption of optimal substructure. That is, it is not sufficient to decompose the power into smaller powers, each of which is computed minimally, since the addition chains for the smaller powers may be related (to share computations). For example, in the shortest addition chain for a^{15} above, the subproblem for a^6 must be computed as $(a^3)^2$ since a^3 is re-used (as opposed to, say, $a^6 = a^2(a^2)^2$, which also requires three multiplies).

Addition-subtraction–Chain Exponentiation

If both multiplication and division are allowed, then an addition-subtraction chain may be used to obtain even fewer total multiplications+divisions (where subtraction corresponds to division). However, the slowness of division compared to multiplication makes this technique unattractive in general. For exponentiation to negative integer powers, on the other hand, since one division is required anyway, an addition-subtraction chain is often beneficial. One such example is a^{-31}, where computing $1/a^{31}$ by a shortest addition chain for a^{31} requires 7 multiplications and one division, whereas the shortest addition-subtraction chain requires 5 multiplications and one division:

$$a^{-31} = a/((((a^2)^2)^2)^2)^2 \text{ (addition-subtraction chain, 5 mults + 1 div).}$$

For exponentiation on elliptic curves, the inverse of a point (x, y) is available at no cost, since it is simply $(x, -y)$, and therefore addition-subtraction chains are optimal in this context even for positive integer exponents.

Arbitrary-precision Arithmetic

In computer science, arbitrary-precision arithmetic, also called bignum arithmetic, multiple-precision arithmetic, or sometimes infinite-precision arithmetic, indicates that calculations are performed on numbers whose digits of precision are limited only by the available memory of the host system. This contrasts with the faster fixed-precision arithmetic found in most arithmetic logic unit (ALU) hardware, which typically offers between 8 and 64 bits of precision.

Several modern programming languages have built-in support for bignums, and others have libraries available for arbitrary-precision integer and floating-point math. Rather than store values as a fixed number of binary bits related to the size of the processor register, these implementations typically use variable-length arrays of digits.

Arbitrary precision is used in applications where the speed of arithmetic is not a limiting factor, or where precise results with very large numbers are required. It should not be confused with the symbolic computation provided by many computer algebra systems, which represent numbers by expressions such as $\pi \cdot \sin(2)$, and can thus *represent* any computable number with infinite precision.

Applications

A common application is public-key cryptography, whose algorithms commonly employ arithmetic with integers having hundreds of digits. Another is in situations where artificial limits and overflows would be inappropriate. It is also useful for checking the results of fixed-precision calculations, and for determining optimal or near-optimal values for coefficients needed in formulae, for example the $\sqrt{1/3}$ that appears in Gaussian integration.

Arbitrary precision arithmetic is also used to compute fundamental mathematical constants such as π to millions or more digits and to analyze the properties of the digit strings or more generally to investigate the precise behaviour of functions such as the Riemann zeta function where certain questions are difficult to explore via analytical methods. Another example is in rendering fractal images with an extremely high magnification, such as those found in the Mandelbrot set.

Arbitrary-precision arithmetic can also be used to avoid overflow, which is an inherent limitation of fixed-precision arithmetic. Similar to a 5-digit odometer's display which changes from 99999 to 00000, a fixed-precision integer may exhibit *wraparound* if numbers grow too large to represent at the fixed level of precision. Some processors can instead deal with overflow by *saturation,* which means that if a result would be unrepresentable, it is replaced with the nearest representable value. (With 16-bit unsigned saturation, adding any positive amount to 65535 would yield 65535.) Some processors can generate an exception if an arithmetic result exceeds the available precision. Where necessary, the exception can be caught and recovered from—for instance, the operation could be restarted in software using arbitrary-precision arithmetic.

In many cases, the task or the programmer can guarantee that the integer values in a specific application will not grow large enough to cause an overflow. Such guarantees may be based on pragmatic limits: a school attendance program may have a task limit of 4,000 students. A programmer may design the computation so that intermediate results stay within specified precision boundaries.

Some programming languages such as Lisp, Python, Perl, Haskell and Ruby use, or have an option to use, arbitrary-precision numbers for *all* integer arithmetic. Although this reduces performance, it eliminates the possibility of incorrect results (or exceptions) due to simple overflow. It also makes it possible to guarantee that arithmetic results will be the same on all machines, regardless of any particular machine's word size. The exclusive use of arbitrary-precision numbers in a programming language also simplifies the language, because *a number is a number* and there is no need for multiple types to represent different levels of precision.

Implementation Issues

Arbitrary-precision arithmetic is considerably slower than arithmetic using numbers that fit entirely within processor registers, since the latter are usually implemented in hardware arithmetic whereas the former must be implemented in software. Even if the computer lacks hardware for certain operations (such as integer division, or all floating-point operations) and software is provided instead, it will use number sizes closely related to the available hardware registers: one or two words only and definitely not N words. There are exceptions, as certain *variable word length* machines of the 1950s and 1960s, notably the IBM 1620, IBM 1401 and the Honeywell *Liberator* series, could manipulate numbers bound only by available storage, with an extra bit that delimited the value.

Numbers can be stored in a fixed-point format, or in a floating-point format as a significand multiplied by an arbitrary exponent. However, since division almost immediately introduces infinitely repeating sequences of digits (such as 4/7 in decimal, or 1/10 in binary), should this possibility arise then either the representation would be truncated at some satisfactory size or else rational numbers would be used: a large integer for the numerator and for the denominator. But even with the greatest common divisor divided out, arithmetic with rational numbers can become unwieldy very quickly: 1/99 – 1/100 = 1/9900, and if 1/101 is then added, the result is 10001/999900.

The size of arbitrary-precision numbers is limited in practice by the total storage available, the variables used to index the digit strings, and computation time. A 32-bit operating system may limit available storage to less than 4 GB. A programming language using 32-bit integers can only index 4 GB. If multiplication is done with a $\Theta(N^2)$ algorithm, it would take on the order of 10^{12} steps to multiply two one-million-word numbers.

Numerous algorithms have been developed to efficiently perform arithmetic operations on numbers stored with arbitrary precision. In particular, supposing that N digits are employed, algorithms have been designed to minimize the asymptotic complexity for large N.

The simplest algorithms are for addition and subtraction, where one simply adds or subtracts the digits in sequence, carrying as necessary, which yields an $O(N)$ algorithm.

Comparison is also very simple. Compare the high-order digits (or machine words) until a difference is found. Comparing the rest of the digits/words is not necessary. The worst case is $\Theta(N)$, but usually it will go much faster.

For multiplication, the most straightforward algorithms used for multiplying numbers by hand (as taught in primary school) require $\Theta(N^2)$ operations, but multiplication algorithms that achieve $O(N \log(N) \log(\log(N)))$ complexity have been devised, such as the Schönhage–Strassen algorithm, based on fast Fourier transforms, and there are also algorithms with slightly worse complexity but with sometimes superior real-world performance for smaller N. The Karatsuba multiplication is such an algorithm.

Pre-set Precision

In some languages such as REXX, the precision of all calculations must be set before doing a calculation. Other languages, such as Python and Ruby extend the precision automatically to prevent overflow.

Example:

The calculation of factorials can easily produce very large numbers. This is not a problem for their usage in many formulae (such as Taylor series) because they appear along with other terms, so that–given careful attention to the order of evaluation–intermediate calculation values are not troublesome. If approximate values of factorial numbers are desired, Stirling's approximation gives good results using floating-point arithmetic. The largest representable value for a fixed-size integer variable may be exceeded even for relatively small arguments as shown in the table. Even floating-point numbers are soon outranged, so it may help to recast the calculations in terms of the logarithm of the number.

But if exact values for large factorials are desired, then special software is required, as in the pseudocode that follows, which implements the classic algorithm to calculate 1, 1×2, 1×2×3, 1×2×3×4, etc. the successive factorial numbers.

```
Constant Limit = 1000;                 % Sufficient digits.

Constant Base = 10;                   % The base of the simulated arithmetic.

Constant FactorialLimit = 365;         % Target number to solve, 365!

Array digit[1:Limit] of integer;       % The big number.

Integer carry,d;                       % Assistants during multiplication.

Integer last,i;                       % Indices to the big number's digits.

Array text[1:Limit] of character; % Scratchpad for the output.

Constant tdigit[0:9] of character = ["0","1","2","3","4","5","6","7",
```

```
"8","9"];

BEGIN

 digit:=0;                          % Clear the whole array.

 digit[1]:=1;                       % The big number starts with 1,

 last:=1;                         % Its highest-order digit is number 1.

 for n:=1 to FactorialLimit do      % Step through producing 1!, 2!, 3!,
4!, etc.

  carry:=0;                         % Start a multiply by n.

  for i:=1 to last do               % Step along every digit.

   d:=digit[i]*n + carry;           % The classic multiply.

   digit[i]:=d mod Base;            % The low-order digit of the result.

   carry:=d div Base;               % The carry to the next digit.

  next i;

  while carry > 0                   % Store the carry in the big number.

   if last >= Limit then croak("Overflow!");  % If possible!

   last:=last + 1;                  % One more digit.

   digit[last]:=carry mod Base;     % Placed.

   carry:=carry div Base;           % The carry reduced.

  Wend                            % With n > Base, maybe > 1 digit extra.

  text:=" ";                        % Now prepare the output.

  for i:=1 to last do               % Translate from binary to text.

   text[Limit - i + 1]:=tdigit[digit[i]];  % Reversing the order.

  next i;                           % Arabic numerals put the low order
last.

  Print text," = ",n,"!";           % Print the result!

 next n;                            % On to the next factorial up.

END;
```

With the example in view, a number of details can be discussed. The most important is the choice of the representation of the big number. In this case, only integer values are required for digits, so an array of fixed-width integers is adequate. It is convenient to have successive elements of the array represent higher powers of the base.

The second most important decision is in the choice of the base of arithmetic, here ten. There are many considerations. The scratchpad variable *d* must be able to hold the result of a single-digit multiply *plus the carry* from the prior digit's multiply. In base ten, a sixteen-bit integer is certainly adequate as it allows up to 32767. However, this example cheats, in that the value of *n* is not itself limited to a single digit. This has the consequence that the method will fail for $n > 3200$ or so. In a more general implementation, *n* would also use a multi-digit representation. A second consequence of the shortcut is that after the multi-digit multiply has been completed, the last value of *carry* may need to be carried into multiple higher-order digits, not just one.

There is also the issue of printing the result in base ten, for human consideration. Because the base is already ten, the result could be shown simply by printing the successive digits of array *digit*, but they would appear with the highest-order digit last (so that 123 would appear as "321"). The whole array could be printed in reverse order, but that would present the number with leading zeroes ("00000...000123") which may not be appreciated, so this implementation builds the representation in a space-padded text variable and then prints that. The first few results (with spacing every fifth digit and annotation added here) are:

Factorial numbers		Reach of computer integers	
1 =	1!		
2 =	2!		
6 =	3!		
24 =	4!		
120 =	5!	8-bit	255
720 =	6!		
5040 =	7!		
40320 =	8!	16-bit	65535
3 62880 =	9!		
36 28800 =	10!		
399 16800 =	11!		
4790 01600 =	12!	32-bit	42949 67295
62270 20800 =	13!		
8 71782 91200 =	14!		
130 76743 68000 =	15!		
2092 27898 88000 =	16!		
35568 74280 96000 =	17!		
6 40237 37057 28000 =	18!		
121 64510 04088 32000 =	19!		
2432 90200 81766 40000 =	20!	64-bit	18446 74407 37095 51615
51090 94217 17094 40000 =	21!		
11 24000 72777 76076 80000 =	22!		

258 52016 73888 49766 40000 =	23!		
6204 48401 73323 94393 60000 =	24!		
1 55112 10043 33098 59840 00000 =	25!		
40 32914 61126 60563 55840 00000 =	26!		
1088 88694 50418 35216 07680 00000 =	27!		
30488 83446 11713 86050 15040 00000 =	28!		
8 84176 19937 39701 95454 36160 00000 =	29!		
265 25285 98121 91058 63630 84800 00000 =	30!		
8222 83865 41779 22817 72556 28800 00000 =	31!		
2 63130 83693 36935 30167 21801 21600 00000 =	32!		
86 83317 61881 18864 95518 19440 12800 00000 =	33!		
2952 32799 03960 41408 47618 60964 35200 00000 =	34!	128-bit	3402 82366 92093 84634 63374 60743 17682 11455
1 03331 47966 38614 49296 66651 33752 32000 00000 =	35!		

This implementation could make more effective use of the computer's built in arithmetic. A simple escalation would be to use base 100 (with corresponding changes to the translation process for output), or, with sufficiently wide computer variables (such as 32-bit integers) we could use larger bases, such as 10,000. Working in a power-of-2 base closer to the computer's built-in integer operations offers advantages, although conversion to a decimal base for output becomes more difficult. On typical modern computers, additions and multiplications take constant time independent of the values of the operands (so long as the operands fit in single machine words), so there are large gains in packing as much of a bignumber as possible into each element of the digit array. The computer may also offer facilities for splitting a product into a digit and carry without requiring the two operations of *mod* and *div* as in the example, and nearly all arithmetic units provide a *carry flag* which can be exploited in multiple-precision addition and subtraction. This sort of detail is the grist of machine-code programmers, and a suitable assembly-language bignumber routine can run much faster than the result of the compilation of a high-level language, which does not provide access to such facilities.

For a single-digit multiply the working variables must be able to hold the value $(\text{base-}1)^2 + \text{carry}$, where the maximum value of the carry is (base-1). Similarly, the variables used to index the digit array are themselves limited in width. A simple way to extend the indices would be to deal with the bignumber's digits in blocks of some convenient size so that the addressing would be via (block *i*, digit *j*) where *i* and *j* would be small integers, or, one could escalate to employing bignumber techniques for the indexing variables. Ultimately, machine storage capacity and execution time impose limits on the problem size.

Software Libraries

Arbitrary-precision arithmetic in most computer software is implemented by calling an external library that provides data types and subroutines to store numbers with the requested precision and to perform computations.

Different libraries have different ways of representing arbitrary-precision numbers, some libraries work only with integer numbers, others store floating point numbers in a variety of bases (decimal or binary powers). Rather than representing a number as single value, some store numbers as a numerator/denominator pair (rationals) and some can fully represent computable numbers, though only up to some storage limit. Fundamentally, Turing machines cannot represent all real numbers, as the cardinality of R exceeds the cardinality of Z.

Exponentiation by Squaring

In mathematics and computer programming, exponentiating by squaring is a general method for fast computation of large positive integer powers of a number, or more generally of an element of a semigroup, like a polynomial or a square matrix. Some variants are commonly referred to as square-and-multiply algorithms or binary exponentiation. These can be of quite general use, for example in modular arithmetic or powering of matrices. For semigroups for which additive notation is commonly used, like elliptic curves used in cryptography, this method is also referred to as double-and-add.

Basic Method

The method is based on the observation that, for a positive integer n, we have:

$$x^n = \begin{cases} x(x^2)^{\frac{n-1}{2}}, & if\, n \text{ is odd} \\ (x^2)^{\frac{n}{2}}, & if\, n \text{ is even.} \end{cases}$$

This method uses the bits of the exponent to determine which powers are computed.

This example shows how to compute x^{13} using this method. The exponent, 13, is 1101 in binary. The bits are used in left to right order. The exponent has 4 bits, so there are 4 iterations.

First, initialize the result to1: $r \leftarrow 1(=x^0)$.

Step 1) $r \leftarrow r^2(=x^0)$; bit 1 = 1, so compute $r \leftarrow r{\cdot}x(=x^1)$;

Step 2) $r \leftarrow r^2(=x^2)$; bit 2 = 1, so compute $r \leftarrow r{\cdot}x(=x^3)$;

Step 3) $r \leftarrow r^2(=x^6)$; bit 3 = 0, so we are done with this step (equivalently, this corresponds to $r \leftarrow r{\cdot}x^0(=x^6)$;

Step 4) $r \leftarrow r^2(=x^{12})$; bit 4 = 1, so compute $r \leftarrow r{\cdot}x(=x^{13})$.

If we write n in binary as $b_k \cdots b_0$, then this is equivalent to defining a sequence $r_{k+1},\ldots,r_0$ by letting $r_{k+1}=1$ and then defining $r_i = r_{i+1}^2 x^{b_i}$ for $i = k,\ldots,0$, where r_0 will equal x^n.

This may be implemented as the following recursive algorithm:

```
Function exp_by_squaring(x, n)

  if n < 0  then return exp_by_squaring(1 / x, -n);

  else if n = 0  then return  1;

  else if n = 1  then return  x ;

  else if n is even  then return exp_by_squaring(x * x,  n / 2);

  else if n is odd  then return x * exp_by_squaring(x * x, (n - 1)
/ 2);
```

Although not tail-recursive, this algorithm may be rewritten into a tail recursive algorithm by introducing an auxiliary function:

```
Function exp_by_squaring(x, n)

  exp_by_squaring2(1, x, n)

Function exp_by_squaring2(y, x, n)

  if n < 0  then return exp_by_squaring2(y, 1 / x, - n);

  else if n = 0  then return  y;

  else if n = 1  then return  x * y;

  else if n is even  then return exp_by_squaring2(y, x * x,  n / 2);

  else if n is odd  then return exp_by_squaring2(x * y, x * x, (n -
1) / 2).
```

The iterative version of the algorithm also uses a bounded auxiliary space, and is given by:

```
Function exp_by_squaring_iterative(x, n)

  if n < 0 then

    x := 1 / x;

    n := -n;

  if n = 0 then return 1

  y := 1;

  while n > 1 do
```

```
    if n is even then
      x := x * x;
      n := n / 2;
    else
      y := x * y;
      x := x * x;
      n := (n - 1) / 2;
  return x * y
```

Computational Complexity

A brief analysis shows that such an algorithm uses $\lfloor \log_2 n \rfloor$ squarings and at most $\lfloor \log_2 n \rfloor$ multiplications, where $\lfloor \ \rfloor$ denotes the floor function. More precisely, the number of multiplications is one less than the number of ones present in the binary expansion of n. For n greater than about 4 this is computationally more efficient than naively multiplying the base with itself repeatedly.

Each squaring results in approximately double the number of digits of the previous, and so, if multiplication of two d-digit numbers is implemented in O(d^k) operations for some fixed k, then the complexity of computing x^n is given by:

$$\sum_{i=0}^{O(\log n)} \left(2^i O(\log x)\right)^k = O\left((n \log x)^k\right).$$

2^k-ary Method

This algorithm calculates the value of x^n after expanding the exponent in base 2^k. It was first proposed by Brauer in 1939. In the algorithm below we make use of the following function $f(0) = (k, 0)$ and $f(m) = (s, u)$, where $m = u \cdot 2^s$ with u odd.

Algorithm:

Input

An element x of G, a parameter $k > 0$, a non-negative integer $n = (n_{l-1}, n_{l-2}, \ldots, n_0)_2{}^k$ and the precomputed values $x^3, x^5, \ldots, x^{2^k-1}$.

Output

The element x^n in G.

```
y := 1; i := l - 1
```

```
while i ≥ 0 do
    (s, u) := f(n_i)
    for j := 1 to k - s do
        y := y^2
    y := y * x^u
    for j := 1 to s do
        y := y^2
    i := i - 1
return y
```

For optimal efficiency, k should be the smallest integer satisfying:

$$\log n < \frac{k(k+1)\cdot 2^{2k}}{2^{k+1}-k-2}+1.$$

Sliding-window Method

This method is an efficient variant of the 2^k-ary method. For example, to calculate the exponent 398, which has binary expansion $(110\ 001\ 110)_2$, we take a window of length 3 using the 2^k-ary method algorithm and calculate 1, x^3, x^6, x^{12}, x^{24}, x^{48}, x^{49}, x^{98}, x^{99}, x^{198}, x^{199}, x^{398}. But, we can also compute 1, x^3, x^6, x^{12}, x^{24}, x^{48}, x^{96}, x^{192}, x^{198}, x^{199}, x^{398}, which saves one multiplication and amounts to evaluating $(110\ 001\ 110)_2$.

Here is the general algorithm:

Algorithm:

Input

An element x of G, a non negative integer $n=(n_{l-1}, n_{l-2}, ..., n_0)_2$, a parameter $k > 0$ and the pre-computed values $x^3, x^5, ..., x^{2^k-1}$.

Output

The element $x^n \in G$.

```
y := 1; i := l - 1
while i > -1 do
    if n_i = 0 then
```

```
        y := y^2, i := i - 1
    else
        s := max{i - k + 1, 0}
        while n_s = 0 do
            s := s + 1
        for h := 1 to i - s + 1 do
            y := y^2
        u := (n_i, n_(i-1), ..., n_s)_2
        y := y * x^u
        i := s - 1
return y
```

Montgomery's Ladder Technique

Many algorithms for exponentiation do not provide defence against side-channel attacks. Namely, an attacker observing the sequence of squarings and multiplications can (partially) recover the exponent involved in the computation. This is a problem if the exponent should remain secret, as with many public-key cryptosystems. A technique called "Montgomery's ladder" addresses this concern.

Given the binary expansion of a positive, non-zero integer $n = (n_{k-1}...n_0)_2$ with $n_{k-1} = 1$, we can compute x^n as follows:

```
x_1 = x; x_2 = x^2
for i = k - 2 to 0 do
  If n_i = 0 then
    x_2 = x_1 * x_2; x_1 = x_1^2
  else
    x_1 = x_1 * x_2; x_2 = x_2^2
return x_1
```

The algorithm performs a fixed sequence of operations (up to log n): a multiplication and squaring takes place for each bit in the exponent, regardless of the bit's specific value. A similar algorithm for multiplication by doubling exists.

This specific implementation of Montgomery's ladder is not yet protected against cache timing attacks: memory access latencies might still be observable to an attacker, as different variables are accessed depending on the value of bits of the secret exponent. Modern cryptographic implementations use a “scatter” technique to make sure the processor always misses the faster cache.

Fixed-base Exponent

There are several methods which can be employed to calculate x^n when the base is fixed and the exponent varies. As one can see, precomputations play a key role in these algorithms.

Yao’s Method

Yao’s method is orthogonal to the 2^k-ary method where the exponent is expanded in radix $b = 2^k$ and the computation is as performed in the algorithm above. Let n, n_i, b, and b_i be integers.

Let the exponent n be written as:

$$n = \sum_{i=0}^{w-1} n_i b_i,$$

where $0 \le n_i < h$ for all $i \in [0, w-1]$.

Let $x_i = x^{b_i}$.

Then the algorithm uses the equality:

$$x^n = \prod_{i=0}^{w-1} x_i^{n_i} = \prod_{j=1}^{h-1} \Big[\prod_{n_i=j} x_i\Big]^j.$$

Given the element x of G, and the exponent n written in the above form, along with the precomputed values $x^{b_0} \ldots x^{b_{w-1}}$, the element x^n is calculated using the algorithm below:

```
y = 1, u = 1, j = h - 1
while j > 0 do
    for i = 0 to w - 1 do
        if n_i = j then
            u = u × x^b_i
    y = y × u
    j = j - 1
return y
```

If we set $h = 2^k$ and $b_i = h^i$, then the n_i values are simply the digits of n in base h. Yao's method collects in u first those x_i that appear to the highest power $h-1$; in the next round those with power $h-2$ are collected in u as well etc. The variable y is multiplied $h-1$ times with the initial u, $h-2$ times with the next highest powers, and so on. The algorithm uses $w+h-2$ multiplications, and $w+1$ elements must be stored to compute x^n.

Euclidean Method

The Euclidean method was first introduced in Efficient exponentiation using precomputation and vector addition chains by P.D Rooij.

This method for computing x^n in group G, where n is a natural integer, whose algorithm is given below, is using the following equality recursively:

$$x_0^{n_0} \cdot x_1^{n_1} = \left(x_0 \cdot x_1^q\right)^{n_0} \cdot x_1^{n_1 \bmod n_0},$$

where $q = \left\lfloor \frac{n_1}{n_0} \right\rfloor$. In other words, a Euclidean division of the exponent n_1 by n_0 is used to return a quotient q and a rest $n_1 \bmod n_0$.

Given the base element x in group G, and the exponent n written as in Yao's method, the element x^n is calculated using l precomputed values $x^{b_0},...,x^{b_{l_i}}$ and then the algorithm below.

```
Begin loop
    Find M∈[0,l−1], such that ∀i∈[0,l−1],n_M ≥ n_i .
    Find N∈([0,l−1]−M), such that ∀i∈([0,l−1]−M),n_N ≥ n_i .
    Break loop if n_N = 0 .
    Let q=⌊n_M / n_N⌋, and then let n_N = (n_M mod n_N) .
    Compute recursively x_M^q, and then let x_N = x_N·x_M^q .
End loop;
Return x^n = x_M^(n_M) .
```

The algorithm first finds the largest value among the n_i and then the supremum within the set of $\{ n_i \setminus i \neq M \}$. Then it raises x_M to the power q, multiplies this value with x_N, and then assigns x_N the result of this computation and n_M the value n_M modulo n_N.

Further Applications

The same idea allows fast computation of large exponents modulo a number. Especially in cryptography, it is useful to compute powers in a ring of integers modulo q. It can also be used to compute integer powers in a group, using the rule:

$$\text{Power}(x, -n) = (\text{Power}(x, n))^{-1}.$$

The method works in every semigroup and is often used to compute powers of matrices.

For example, the evaluation of:

$$13789^{722341} \pmod{2345} = 2029$$

would take a very long time and lots of storage space if the naïve method were used: compute 13789^{722341}, then take the remainder when divided by 2345. Even using a more effective method will take a long time: square 13789, take the remainder when divided by 2345, multiply the result by 13789, and so on. This will take less than $2\log_2(722340) \leq 40$ modular multiplications.

Applying above *exp-by-squaring* algorithm, with "*" interpreted as $x * y = xy$ mod 2345 (that is, a multiplication followed by a division with remainder) leads to only 27 multiplications and divisions of integers, which may all be stored in a single machine word.

Example Implementations

Computation by Powers of 2

This is a non-recursive implementation of the above algorithm in Ruby. `n = n-1` is redundant when `n = n / 2` implicitly rounds towards zero, as strongly-typed languages with integer division would do. (`n = n << 1` has the same effect.) `n[0]` is the rightmost bit of the binary representation of `n`, so if it is 1, then the number is odd, and if it is zero, then the number is even. It is also `n` modulo 2. (`n & 1` has the same effect.)

```
def power(x, n)

  result = 1

  while n.nonzero?

    if n[0].nonzero?

      result *= x

      n -= 1

    end

    x *= x

    n /= 2
```

```
  end
  return result
end
```

Runtime Example: Compute 3^{10}

```
parameter x =  3
parameter n = 10
result := 1

Iteration 1
  n = 10 -> n is even
  x := x² = 3² = 9
  n := n / 2 = 5

Iteration 2
  n = 5 -> n is odd
      -> result := result * x = 1 * x = 1 * 3² = 9
         n := n - 1 = 4
  x := x² = 9² = 3⁴ = 81
  n := n / 2 = 2

Iteration 3
  n = 2 -> n is even
  x := x² = 81² = 3⁸ = 6561
  n := n / 2 = 1

Iteration 4
  n = 1 -> n is odd
      -> result := result * x = 3² * 3⁸ = 3¹⁰ = 9 * 6561 = 59049
```

```
        n := n - 1 = 0

return result

result := 3

bin := "1010"

Iteration for digit 4:

  result := result² = 3² = 9

  1010bin - Digit equals "0"

Iteration for digit 3:

  result := result² = (3²)² = 3⁴  = 81

  1010bin - Digit equals "1" --> result := result*3 = (3²)²*3 = 3⁵  = 243

Iteration for digit 2:

  result := result² = ((3²)²*3)² = 3¹⁰  = 59049

  1010bin - Digit equals "0"

return result
```

This example is based on the algorithm above. If calculated by hand, should go from left to right. If the start number is 1, just ignore it. Then if the next is one, square and multiply. If the next is zero, only square.

Calculation of Products of Powers

Exponentiation by squaring may also be used to calculate the product of 2 or more powers. If the underlying group or semigroup is commutative, then it is often possible to reduce the number of multiplications by computing the product simultaneously.

Example:

The formula $a^7 \times b^5$ may be calculated within 3 steps:

$((a)^2 \times a)^2 \times a$ (4 multiplications for calculating a^7),

$((b)^2)^2 \times b$ (3 multiplications for calculating b^5),

$(a^7) \times (b^5)$ (1 multiplication to calculate the product of the two),

so one gets 8 multiplications in total.

A faster solution is to calculate both powers simultaneously:

$$((a \times b)^2 \times a)^2 \times a \times b,$$

which needs only 6 multiplications in total. Note that $a \times b$ is calculated twice; the result could be stored after the first calculation, which reduces the count of multiplication to 5.

Example with numbers:

$$2^7 \times 3^5 = ((2 \times 3)^2 \times 2)^2 \times 2 \times 3 = (6^2 \times 2)^2 \times 6 = 72^2 \times 6 = 31104.$$

Calculating the powers simultaneously instead of calculating them separately always reduces the count of multiplications if at least two of the exponents are greater than 1.

Using Transformation

The example above $a^7 \times b^5$ may also be calculated with only 5 multiplications if the expression is transformed before calculation:

$a^7 \times b^5 = a^2 \times (ab)^5$, with $ab := a \times b$,

$ab := a \times b$ (1 multiplication),

$a^2 \times (ab)^5 = ((ab)^2 \times a)^2 \times ab$ (4 multiplications).

Generalization of transformation shows the following scheme:

For calculating $a^A \times b^B \times \ldots \times m^M \times n^N$

- Define $ab := a \times b$, $abc = ab \times c$, ...
- Calculate the transformed expression $a^{A-B} \times ab^{B-C} \times \ldots \times abc..m^{M-N} \times abc..mn^N$.

Transformation before calculation often reduces the count of multiplications, but in some cases it also increases the count, so it may be a good idea to check the count of multiplications before using the transformed expression for calculation.

Examples:

For the following expressions the count of multiplications is shown for calculating each power separately, calculating them simultaneously without transformation, and calculating them simultaneously after transformation.

Example	$a^7 \times b^5 \times c^3$	$a^5 \times b^5 \times c^3$	$a^7 \times b^4 \times c^1$
Separate	$[((a)^2 \times a)^2 \times a] \times [((b)^2)^2 \times b] \times [(c)^2 \times c]$ (11 multiplications)	$[((a)^2)^2 \times a] \times [((b)^2)^2 \times b] \times [(c)^2 \times c]$ (10 multiplications)	$[((a)^2 \times a)^2 \times a] \times [((b)^2)^2] \times [c]$ (8 multiplications)
Simultaneous	$((a \times b)^2 \times a \times c)^2 \times a \times b \times c$ (8 multiplications)	$((a \times b)^2 \times c)^2 \times a \times b \times c$ (7 multiplications)	$((a \times b)^2 \times a)^2 \times a \times c$ (6 multiplications)
Transformation	a := 2 ab := a×b abc := ab×c (2 multiplications)	a := 2 ab := a×b abc := ab×c (2 multiplications)	a := 2 ab := a×b abc := ab×c (2 multiplications)
Calculation after that	$(a \times ab \times abc)^2 \times abc$ (4 multiplications ⇒ 6 in total)	$(ab \times abc)^2 \times abc$ (3 multiplications ⇒ 5 in total)	$(a \times ab)^2 \times a \times ab \times abc$ (5 multiplications ⇒ 7 in total)

Signed-digit Recoding

In certain computations it may be more efficient to allow negative coefficients and hence use the inverse of the base, provided inversion in G is "fast" or has been precomputed. For example, when computing x^{2^k-1}, the binary method requires $k-1$ multiplications and $k-1$ squarings. However, one could perform k squarings to get x^{2^k} and then multiply by x^{-1} to obtain x^{2^k-1}.

To this end we define the signed-digit representation of an integer n in radix b as:

$$n = \sum_{i=0}^{l-1} n_i b^i \ \ with \ \ |n_i| < b.$$

Signed binary representation corresponds to the particular choice $b = 2$ and $n_i \in \{-1,0,1\}$.. It is denoted by $(n_{l-1} \ldots n_0)_s$. There are several methods for computing this representation. The representation is not unique. For example, take $n = 478$: two distinct signed-binary representations are given by $(10\bar{1}1100\bar{1}10)_s$ and $(100\bar{1}1000\bar{1}0)_s$, where $\bar{1}$ is used to denote -1. Since the binary method computes a multiplication for every non-zero entry in the base-2 representation of n, we are interested in finding the signed-binary representation with the smallest number of non-zero entries, that is, the one with *minimal* Hamming weight. One method of doing this is to compute the representation in non-adjacent form, or NAF for short, which is one that satisfies $n_i n_{i+1} = 0$ for all $i \geq 0$ and denoted by $(n_{l-1} \ldots n_0)_{NAF}$. For example, the NAF representation of 478 is $(1000\bar{1}000\bar{1}0)_{NAF}$. This representation always has minimal Hamming weight. A simple algorithm to compute the NAF representation of a given integer $n = (n_l n_{l-1} \ldots n_0)_2$ with $n_l = n_{l-1} = 0$ is the following:

$$c_0 = 0$$

for $i = 0$ to $l - 1$ do:

$$c_{i+1} = \left\lfloor \frac{1}{2}(c_i + n_i + n_{i+1}) \right\rfloor$$

$$n'_i = c_i + n_i - 2c_{i+1}$$

return $(n'_{l-1} \ldots n'_0)_{\text{NAF}}$.

Another algorithm by Koyama and Tsuruoka does not require the condition that $n_i = n_{i+1} = 0$; it still minimizes the Hamming weight.

Alternatives and Generalizations

Exponentiation by squaring can be viewed as a suboptimal addition-chain exponentiation algorithm: it computes the exponent by an addition chain consisting of repeated exponent doublings (squarings) and/or incrementing exponents by *one* (multiplying by *x*) only. More generally, if one allows *any* previously computed exponents to be summed (by multiplying those powers of *x*), one can sometimes perform the exponentiation using fewer multiplications (but typically using more memory). The smallest power where this occurs is for $n = 15$:

$$x^{15} = x \times (x \times [x \times x^2]^2)^2 \quad \text{(squaring, 6 multiplies)},$$

$$x^{15} = x^3 \times ([x^3]^2)^2 \quad \text{(optimal addition chain, 5 multiplies if } x^3 \text{ is re-used)}.$$

In general, finding the *optimal* addition chain for a given exponent is a hard problem, for which no efficient algorithms are known, so optimal chains are typically only used for small exponents (e.g. in compilers where the chains for small powers have been pre-tabulated). However, there are a number of heuristic algorithms that, while not being optimal, have fewer multiplications than exponentiation by squaring at the cost of additional bookkeeping work and memory usage. Regardless, the number of multiplications never grows more slowly than $\Theta(\log n)$, so these algorithms only improve asymptotically upon exponentiation by squaring by a constant factor at best.

Methods of Computing Square Roots

Methods of computing square roots are numerical analysis algorithms for finding the principal, or non-negative, square root (usually denoted $\sqrt{S}, \sqrt[2]{S}$, or $S^{1/2}$) of a real number. Arithmetically, it means given S, a procedure for finding a number which when multiplied by itself, yields S; algebraically, it means a procedure for finding the non-negative root of the equation $x^2 - S = 0$; geometrically, it means given the area of a square, a procedure for constructing a side of the square.

Every real number has two square roots. The principal square root of most numbers is an irrational number or repeating decimal, so that any computed result is a finite approximation to the square root. For some numbers, the square root is finitely representable in the base (i.e. decimal or binary); for these numbers, including integers known as perfect squares, the square root is exact. However, even if the square root is exact, the procedure used to compute it may return an approximation.

The most common analytical methods are iterative and consist of two steps: finding a suitable starting value, followed by iterative refinement until some termination criteria is met. The starting value can be any number, but fewer iterations will be required the closer it is to the final result. The most familiar such method, most suited for programmatic calculation, is Newton's method, which is based on a property of the derivative in the calculus. A few methods like paper-and-pencil synthetic division and series expansion, do not require a starting value. In some applications, an integer square root is required, which is the square root rounded or truncated to the nearest integer (a modified procedure may be employed in this case).

The method employed depends on what the result is to be used for (i.e. how accurate it has to be), how much effort one is willing to put into the procedure, and what tools are at hand. The methods may be roughly classified as those suitable for mental calculation, those usually requiring at least paper and pencil, and those which are implemented as programs to be executed on a digital electronic computer or other computing device. Algorithms may take into account convergence (how many iterations are required to achieve a specified precision), computational complexity of individual operations (i.e. division) or iterations, and error propagation (the accuracy of the final result).

Procedures for finding square roots (particularly the square root of 2) have been known since at least the period of ancient Babylon in the 17th century BCE. Heron's method from first century Egypt was the first ascertainable algorithm for computing square root. Modern analytic methods began to be developed after introduction of the Arabic numeral system to western Europe in the early Renaissance. Today, nearly all computing devices have a fast and accurate square root function, either as a programming language construct, a compiler intrinsic or library function, or as a hardware operator, based on one of the described procedures.

Initial Estimate

Many iterative square root algorithms require an initial seed value. The seed must be a non-zero positive number; it should be between 1 and S, the number whose square root is desired, because the square root must be in that range. If the seed is far away from the root, the algorithm will require more iterations. If one initializes with $x_0 = 1$ (or S), then approximately $\frac{1}{2}|\log_2 S|$ iterations will be wasted just getting the order of magnitude of the root. It is therefore useful to have a rough estimate, which may have limited accuracy but is easy to calculate. In general, the better the initial estimate, the faster the convergence. For Newton's method (also called Babylonian or Heron's method), a seed somewhat larger than the root will converge slightly faster than a seed somewhat smaller than the root.

In general, an estimate is pursuant to an arbitrary interval known to contain the root (such as $[x_0, 1/x_0]$). The estimate is a specific value of a functional approximation to $f(x) = \sqrt{x}$ over the interval. Obtaining a better estimate involves either obtaining tighter bounds on the interval, or finding a better functional approximation to f(x). The latter usually means using a higher order polynomial in the approximation, though not all approximations are polynomial. Common methods of estimating include scalar, linear, hyperbolic and logarithmic. A decimal base is usually used for mental or paper-and-pencil estimating. A binary base is more suitable for computer estimates. In estimating, the exponent and mantissa are usually treated separately, as the number would be expressed in scientific notation.

Decimal Estimates

Typically the number S is expressed in scientific notation as $a \times 10^{2n}$ where $1 \leq a < 100$ and n is an integer, and the range of possible square roots is $\sqrt{\ } \times 10$ where $1 \leq \sqrt{a} < 10$.

Scalar Estimates

Scalar methods divide the range into intervals, and the estimate in each interval is represented by a single scalar number. If the range is considered as a single interval, the arithmetic mean (5) or geometric mean ($\sqrt{10} \approx 3.16$) times 10^n are plausible estimates. The absolute and relative error for these will differ. In general, a single scalar will be very inaccurate. Better estimates divide the range into two or more intervals, but scalar estimates have inherently low accuracy.

For two intervals, divided geometrically, the square root $\sqrt{S} = \sqrt{a} \times 10^n$ can be estimated as:

$$\sqrt{S} \approx \begin{cases} 2 \cdot 10^n & if\, a < 10, \\ 6 \cdot 10^n & if\, a \geq 10. \end{cases}$$

This estimate has maximum absolute error of $4 \cdot 10^n$ at a = 100, and maximum relative error of 100% at a = 1.

For example, for $S = 125348$ factored as 12.5348×10^4, the estimate is $\sqrt{S} \approx 6 \cdot 10^2 = 600$. $\sqrt{125348} = 354.0$, an absolute error of 246 and relative error of almost 70%.

Linear Estimates

A better estimate, and the standard method used, is a linear approximation to the function $y = x^2$ over a small arc. If, as above, powers of the base are factored out of the number S and the interval reduced to [1,100], a secant line spanning the arc, or a tangent line somewhere along the arc may be used as the approximation, but a least-squares regression line intersecting the arc will be more accurate.

A least-squares regression line minimizes the average difference between the estimate and the value of the function. Its equation is $y = 8.7x - 10$. Reordering, $x = 0.115y + 1.15$. Rounding the coefficients for ease of computation,

$$\sqrt{S} \approx (a/10 + 1.2) \cdot 10^n$$

That is the best estimate *on average* that can be achieved with a single piece linear approximation of the function y=x² in the interval [1,100]. It has a maximum absolute error of 1.2 at a=100, and maximum relative error of 30% at S=1 and 10.

To divide by 10, subtract one from the exponent of a, or figuratively move the decimal point one digit to the left. For this formulation, any additive constant 1 plus a small increment will make a satisfactory estimate so remembering the exact number isn't a burden. The approximation (rounded or not) using a single line spanning the range [1,100] is less than one significant digit of precision; the relative error is greater than $1/2^2$, so less than 2 bits of information are provided. The accuracy is severely limited because the range is two orders of magnitude, quite large for this kind of estimation.

A much better estimate can be obtained by a piece-wise linear approximation: multiple line segments, each approximating some subarc of the original. The more line segments used, the better the approximation. The most common way is to use tangent lines; the critical choices are how to divide the arc and where to place the tangent points. An efficacious way to divide the arc from y=1 to y=100 is geometrically: for two intervals, the bounds of the intervals are the square root of the bounds of the original interval, 1*100, i.e. $\left[1, \sqrt[2]{100}\right]$, and $\left[\sqrt[2]{100}, 100\right]$. For three intervals, the bounds are the cube roots of 100: $\left[1, \sqrt[3]{100}\right]$, $\left[\sqrt[3]{100}\left(\sqrt[3]{100}\right)^2\right]$ and $\left[\left(\sqrt[3]{100}\right)^2, 100\right]$, etc. For two intervals, $\sqrt[2]{100} = 10$,, a very convenient number. Tangent lines are easy to derive, and are located at $x = \sqrt{1 * \sqrt{10}}$ and $x = \sqrt{10 * \sqrt{10}}$. Their equations are: y = 3.56x - 3.16 and y = 11.2x - 31.6. Inverting, the square roots are: x = 0.28y + 0.89 and x = .089y + 2.8. Thus for S = a * 10^{2n}:

$$\sqrt{S} \approx \begin{cases} (0.28a + 0.89) \cdot 10^n & if\, a < 10, \\ (.089a + 2.8) \cdot 10^n & if\, a \geq 10. \end{cases}$$

The maximum absolute errors occur at the high points of the intervals, at a=10 and 100, and are 0.54 and 1.7 respectively. The maximum relative errors are at the endpoints of the intervals, at a=1, 10 and 100, and are 17% in both cases. 17% or 0.17 is larger than 1/10, so the method yields less than a decimal digit of accuracy.

Hyperbolic Estimates

In some cases, hyperbolic estimates may be efficacious, because a hyperbola is also a convex curve and may lie along an arc of $Y = x^2$ better than a line. Hyperbolic estimates are more computationally complex, because they necessarily require a floating division. A near-optimal hyperbolic approximation to x^2 on the interval [1,100] is y=190/(10-x)-20. Transposing, the square root is x = -190/(y+20)+10. Thus for $S = a{\cdot}10^{2n}$:

$$\sqrt{S} \approx (\frac{-190}{a+20}+10){\cdot}10^n$$

The floating division need be accurate to only one decimal digit, because the estimate overall is only that accurate, and can be done mentally. A hyperbolic estimate is better on average than scalar or linear estimates. It has maximum absolute error of 1.58 at 100 and maximum relative error of 16.0% at 10. For the worst case at a=10, the estimate is 3.67. If one starts with 10 and applies Newton-Raphson iterations straight away, two iterations will be required, yielding 3.66, before the accuracy of the hyperbolic estimate is exceeded. For a more typical case like 75, the hyperbolic estimate is 8.00, and 5 Newton-Raphson iterations starting at 75 would be required to obtain a more accurate result.

Arithmetic Estimates

A method analogous to piece-wise linear approximation but using only arithmetic instead of algebraic equations, uses the multiplication tables in reverse: the square root of a number between 1 and 100 is between 1 and 10, so if we know 25 is a perfect square (5x5), and 36 is a perfect square (6x6), then the square root of a number greater than or equal to 25 but less than 36, begins with a 5. Similarly for numbers between other squares. This method will yield a correct first digit, but it is not accurate to one digit: the first digit of the square root of 35 for example, is 5, but the square root of 35 is almost 6.

A better way is to the divide the range into intervals half way between the squares. So any number between 25 and half way to 36, which is 30.5, estimate 5; any number greater than 30.5 up to 36, estimate 6. The procedure only requires a little arithmetic to find a boundary number in the middle of two products from the multiplication table. Here is a reference table of those boundaries:

a	Nearest square	$k = \sqrt{a}$ est.
1 - 2.5	1 (1^2)	1
2.5 - 6.5	4 (2^2)	2

6.5 - 12.5	9 (3^2)	3
12.5 - 20.5	16 (4^2)	4
20.5 - 30.5	25 (5^2)	5
30.5 - 42.5	36 (6^2)	6
42.5 - 56.5	49 (7^2)	7
56.5 - 72.5	64 (8^2)	8
72.5 - 90.5	81 (9^2)	9
90.5 - 100	100 (10^2)	10

The final operation is to multiply the estimate k by the power of ten divided by 2, so for $S = a{\cdot}10^{2n}$,

$$\sqrt{S} \approx k{\cdot}10^n$$

The method implicitly yields one significant digit of accuracy, since it rounds to the best first digit.

The method can be extended 3 significant digits in most cases, by interpolating between the nearest squares bounding the operand. If $k^2 \le a < (k+1)^2$, then $\sqrt{a}$ is approximately k plus a fraction, the difference between a and k² divided by the difference between the two squares:

$$\sqrt{a} \approx k + R$$

where,

$$R = \frac{(a - k^2)}{(k+1)^2 - k^2}$$

The final operation, as above, is to multiply the result by the power of ten divided by 2;

$$\sqrt{S} = \sqrt{a}{\cdot}10^n \approx (k+R){\cdot}10^n$$

k is a decimal digit and R is a fraction that must be converted to decimal. It usually has only a single digit in the numerator, and one or two digits in the denominator, so the conversion to decimal can be done mentally.

Example: find the square root of 75. 75 = 75 * $10^{2 \cdot 0}$, so a is 75 and n is 0. From the multiplication tables, the square root of the mantissa must be 8 point *something* because 8x8 is 64, but 9x9 is 81, too big, so k is 8; *something* is the decimal representation of R. The fraction R is 75-k² = 11, the numerator, and 81-k² = 17, the denominator. 11/17 is a little less than 12/18, which is 2/3s or .67, so guess .66 (it's ok to guess here, the error is very small). So the estimate is 8+.66=8.66. $\sqrt{75}$ to three significant digits is 8.66, so the estmate is good to 3 significant digits. Not all such estimates using this method will be so accurate, but they will be close.

Binary Estimates

When working in the binary numeral system (as computers do internally), by expressing S as $a \times 2^{2n}$ where $0.1_2 \le a < 10_2$, the square root $\sqrt{S} = \sqrt{a} \times 2^n$ can be estimated as:

$$\sqrt{S} \approx (0.485 + 0.485 \cdot a) \cdot 2^n$$

which is the least-squares regression line to 3 significant digit coefficients. $\sqrt{a}$ has maximum absolute error of 0.0408 at a=2, and maximum relative error of 3.0% at a=1. A computationally convenient rounded estimate (because the coefficients are powers of 2) is:

$$\sqrt{S} \approx (0.5 + 0.5 \cdot a) \cdot 2^n$$

which has maximum absolute error of 0.086 at 2 and maximum relative error of 6.1% at a =0.5 and a =2.0.

For $S = 125348 = 1\,1110\,1001\,1010\,0100_2 = 1.1110\,1001\,1010\,0100_2 \times 2^{16}$ the binary approximation gives

$$\sqrt{S} \approx (0.5 + 0.5 \cdot a) \cdot 2^8 = 1.0111\,0100\,1101\,0010_2 \cdot 1\,0000\,0000_2 = 1.456 \cdot 256 = 372.8$$

$$\sqrt{125348} = 354.0,$$

so the estimate has an absolute error of 19 and relative error of 5.3%. The relative error is a little less than $1/2^4$, so the estimate is good to 4+ bits.

An estimate for a good to 8 bits can be obtained by table lookup on the high 8 bits of a, remembering that the high bit is implicit in most floating point representations, and the bottom bit of the 8 should be rounded. The table is 256 bytes of precomputed 8-bit square root values. For example, for the index 11101101_2 representing 1.8515625_{10}, the entry is 10101110_2 representing 1.359375_{10}, the square root of 1.8515625_{10} to 8 bit precision (2+ decimal digits).

Babylonian Method

Perhaps the first algorithm used for approximating $\sqrt{S}$ is known as the Babylonian method, despite there being no direct evidence, beyond informed conjecture, that the eponymous Babylonian mathematicians employed exactly this method. The method is also known as Heron's method, after the first-century Greek mathematician Hero of Alexandria who gave the first explicit description of the method in his AD 60 work Metrica. The basic idea is that if x is an overestimate to the square root of a non-negative real number S then S/x will be an underestimate, or vice versa, and so the average of these two numbers may reasonably be expected to provide a better approximation (though the formal proof of that assertion depends on the inequality of arithmetic and geometric means that shows this average is always an overestimate of the square root). This is equivalent to using Newton's method to solve $x^2 - S = 0$.

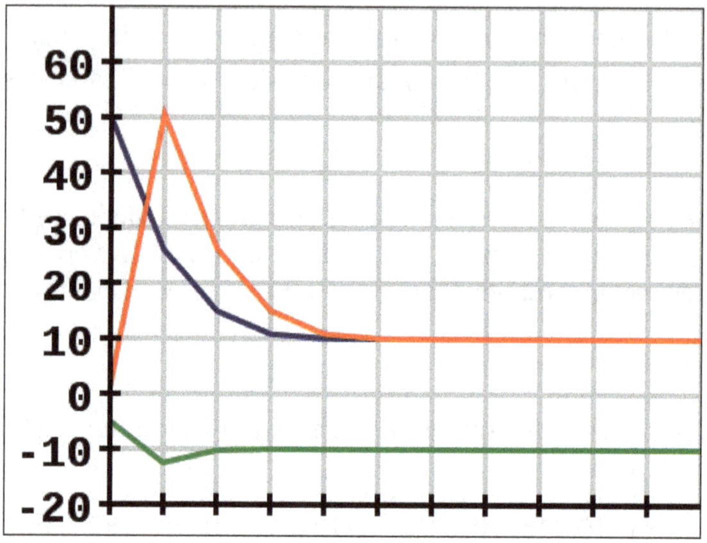

Graph charting the use of the Babylonian method for approximating a square root of 100 (±10) using starting values $x_0 = 50$, $x_0 = 1$, and $x_0 = -5$. Note that a positive starting value yields the positive root, and a negative starting value the negative root.

More precisely, if x is our initial guess of $\sqrt{S}$ and e is the error in our estimate such that $S = (x+e)^2$, then we can expand the binomial and solve for:

$$e = \frac{S - x^2}{2x + e} \approx \frac{S - x^2}{2x},$$

since,

$$e \ll x.$$

Therefore, we can compensate for the error and update our old estimate as:

$$x + e \approx x + \frac{S - x^2}{2x} = \frac{S + x^2}{2x} = \frac{\frac{S}{x} + x}{2} \equiv x_{revised}$$

Since the computed error was not exact, this becomes our next best guess. The process of updating is iterated until desired accuracy is obtained. This is a quadratically convergent algorithm, which means that the number of correct digits of the approximation roughly doubles with each iteration. It proceeds as follows:

- Begin with an arbitrary positive starting value x_0 (the closer to the actual square root of S, the better).

- Let x_{n+1} be the average of x_n and $\frac{S}{x_n}$ (using the arithmetic mean to approximate the geometric mean).
- Repeat step 2 until the desired accuracy is achieved.

It can also be represented as:

$$x_0 \approx \sqrt{S},$$

$$x_{n+1} = \frac{1}{2}\left(x_n + \frac{S}{x_n}\right),$$

$$\sqrt{S} = \lim_{n\to\infty} x_n.$$

This algorithm works equally well in the p-adic numbers, but cannot be used to identify real square roots with p-adic square roots; one can, for example, construct a sequence of rational numbers by this method that converges to +3 in the reals, but to −3 in the 2-adics.

Example:

To calculate $\sqrt{S}$, where $S = 125348$, to six significant figures, use the rough estimation method above to get:

$$\begin{aligned}
x_0 &= 6\cdot 10^2 & & &= 600.000\\
x_1 &= \frac{1}{2}\left(x_0 + \frac{S}{x_0}\right) &= \frac{1}{2}\left(600.000 + \frac{125348}{600.000}\right) &&= 404.457\\
x_2 &= \frac{1}{2}\left(x_1 + \frac{S}{x_1}\right) &= \frac{1}{2}\left(404.457 + \frac{125348}{404.457}\right) &&= 357.187\\
x_3 &= \frac{1}{2}\left(x_2 + \frac{S}{x_2}\right) &= \frac{1}{2}\left(357.187 + \frac{125348}{357.187}\right) &&= 354.059\\
x_4 &= \frac{1}{2}\left(x_3 + \frac{S}{x_3}\right) &= \frac{1}{2}\left(354.059 + \frac{125348}{354.059}\right) &&= 354.045\\
x_5 &= \frac{1}{2}\left(x_4 + \frac{S}{x_4}\right) &= \frac{1}{2}\left(354.045 + \frac{125348}{354.045}\right) &&= 354.045
\end{aligned}$$

Therefore,

$$\sqrt{125348} \approx 354.045.$$

Convergence

Suppose that $x_0 > 0$ and $S > 0$. Then for any natural number n, $x_n > 0$. Let the relative error in x_n be defined by:

$$\varepsilon_n = \frac{x_n}{\sqrt{S}} - 1 > -1$$

and thus,

$$\varepsilon_{n+1} = \frac{\varepsilon_n^2}{2(1+\varepsilon_n)} \geq 0.$$

And thus that,

$$\varepsilon_{n+2} \leq \min\left\{\frac{\varepsilon_{n+1}^2}{2}, \frac{\varepsilon_{n+1}}{2}\right\}$$

and consequently that convergence is assured, and quadratic.

Worst Case for Convergence

If using the rough estimate above with the Babylonian method, then the least accurate cases in ascending order are as follows:

$$\begin{aligned}
S &= 1; & x_0 &= 2; & x_1 &1.250; & \varepsilon_1 &= 0.250. \\
S &= 10; & x_0 &= 2; & x_1 &= 3.500; & \varepsilon_1 &< 0.107. \\
S &= 10; & x_0 &= 6; & x_1 &= 11.333; & \varepsilon_1 &< 0.213. \\
S &= 100; & x_0 &= 6; & x_1 &= 11.333; & \varepsilon_1 &< 0.134.
\end{aligned}$$

Thus in any case,

$$\begin{aligned}
&\varepsilon_1 \leq 2^{-2}. \\
&\varepsilon_2 < 2^{-5} < 10^{-1}. \\
&\varepsilon_3 < 2^{-11} < 10^{-3}. \\
&\varepsilon_4 < 2^{-23} < 10^{-6}. \\
&\varepsilon_5 < 2^{-47} < 10^{-14}. \\
&\varepsilon_6 < 2^{-95} < 10^{-28}. \\
&\varepsilon_7 < 2^{-191} < 10^{-57}. \\
&\varepsilon_8 < 2^{-383} < 10^{-115}.
\end{aligned}$$

Rounding errors will slow the convergence. It is recommended to keep at least one extra digit beyond the desired accuracy of the x_n being calculated to minimize round off error.

Bakhshali Method

This method for finding an approximation to a square root was described in an ancient Indian mathematical manuscript called the Bakhshali manuscript. It is equivalent to two iterations of the Babylonian method beginning with x_0. Thus, the algorithm is quartically convergent, which means that the number of correct digits of the approximation roughly quadruples with each iteration. The original presentation, using modern notation, is as follows: To calculate $\sqrt{S}$, let x_0^2 be the initial approximation to S. Then, successively iterate as:

$$a_n = \frac{S - x_n^2}{2x_n},$$

$$b_n = x_n + a_n,$$

$$x_{n+1} = b_n - \frac{a_n^2}{2b_n}.$$

Written explicitly, it becomes:

$$x_{n+1} = (x_n + a_n) - \frac{a_n^2}{2(x_n + a_n)}.$$

Let $x_0 = N$ be an integer which is the nearest perfect square to S. Also, let the difference $d = S - N^2$, then the first iteration can be written as:

$$\sqrt{S} \approx N + \frac{d}{2N} - \frac{d^2}{8N^3 + 4Nd} = \frac{8N^4 + 8N^2 d + d^2}{8N^3 + 4Nd} = \frac{N^4 + 6N^2 S + S^2}{4N^3 + 4NS} = \frac{N^2(N^2 + 6S) + S^2}{4N(N^2 + S)}.$$

This gives a rational approximation to the square root.

Example:

Using the same example as given in Babylonian method, let $S = 125348$. Then, the first iterations gives:

$$x_0 = 600$$

$$a_1 = \frac{125348 - 600^2}{2 \times 600} = -195.543$$

$$x_1 = 404.456 - \frac{(-195.543)^2}{2 \times 404.456} = 357.186$$

Likewise the second iteration gives:

$$a_2 = \frac{125348 - 357.186^2}{2 \times 357.186} = -3.126$$

$$b_2 = 357.186 + (-3.126) = 354.06$$

$$x_2 = 354.06 - \frac{(-3.1269)^2}{2 \times 354.06} = 354.046$$

Digit-by-digit Calculation

This is a method to find each digit of the square root in a sequence. It is slower than the Babylonian method, but it has several advantages:

- It can be easier for manual calculations.
- Every digit of the root found is known to be correct, i.e., it does not have to be changed later.
- If the square root has an expansion that terminates, the algorithm terminates after the last digit is found. Thus, it can be used to check whether a given integer is a square number.
- The algorithm works for any base, and naturally, the way it proceeds depends on the base chosen.

Napier's bones include an aid for the execution of this algorithm. The shifting *n*th root algorithm is a generalization of this method.

Basic Principle

First, consider the case of finding the square root of a number *Z*, that is the square of a two-digit number *XY*, where *X* is the tens digit and *Y* is the units digit. Specifically:

$$Z = (10X + Y)^2 = 100X^2 + 20XY + Y^2$$

Now using the Digit-by-Digit algorithm, we first determine the value of *X*. *X* is the largest digit such that X^2 is less or equal to *Z* from which we removed the two rightmost digits.

In the next iteration, we pair the digits, multiply *X* by 2, and place it in the tenth's place while we try to figure out what the value of *Y* is.

Since this is a simple case where the answer is a perfect square root *XY*, the algorithm stops here.

The same idea can be extended to any arbitrary square root computation next. Suppose we are able to find the square root of N by expressing it as a sum of n positive numbers such that:

$$N = (a_1 + a_2 + a_3 + \cdots + a_n)^2.$$

By repeatedly applying the basic identity:

$$(x+y)^2 = x^2 + 2xy + y^2,$$

the right-hand-side term can be expanded as:

$$\begin{aligned}
&(a_1 + a_2 + a_3 + \cdots + a_n)^2 \\
&= a_1^2 + 2a_1a_2 + a_2^2 + 2(a_1 + a_2)a_3 + a_3^2 + \cdots + a_{n-1}^2 + 2\left(\sum_{i=1}^{n-1} a_i\right)a_n + a_n^2 \\
&= a_1^2 + [2a_1 + a_2]a_2 + [2(a_1 + a_2) + a_3]a_3 + \cdots + \left[2\left(\sum_{i=1}^{n-1} a_i\right) + a_n\right]a_n.
\end{aligned}$$

This expression allows us to find the square root by sequentially guessing the values of a_is. Suppose that the numbers $a_1, \ldots, a_{m-1}$ have already been guessed, then the m-th term of the right-hand-side of above summation is given by $Y_m = [2P_{m-1} + a_m]a_m$, where $P_{m-1} = \sum_{i=1}^{m-1} a_i$ is the approximate square root found so far. Now each new guess a_m should satisfy the recursion:

$$X_m = X_{m-1} - Y_m,$$

such that $X_m \geq 0$ for all $1 \leq m \leq n$, with initialization $X_0 = N$. When $X_n = 0$, the exact square root has been found; if not, then the sum of a_is gives a suitable approximation of the square root, with X_n being the approximation error.

For example, in the decimal number system we have:

$$N = (a_1 \cdot 10^{n-1} + a_2 \cdot 10^{n-2} + \cdots + a_{n-1} \cdot 10 + a_n)^2,$$

where 10^{n-i} are place holders and the coefficients $a_i \in \{0,1,2,\ldots,9\}$. At any m-th stage of the square root calculation, the approximate root found so far, Y_m and the summation term are given by:

$$P_{m-1} = \sum_{i=1}^{m-1} a_i \cdot 10^{n-i} = 10^{n-m+1} \sum_{i=1}^{m-1} a_i \cdot 10^{m-i-1},$$

$$Y_m = [2P_{m-1} + a_m \cdot 10^{n-m}]a_m \cdot 10^{n-m} = [20\sum_{i=1}^{m-1} a_i \cdot 10^{m-i-1} + a_m]a_m \cdot 10^{2(n-m)}.$$

Here since the place value of Y_m is an even power of 10, we only need to work with the pair of most significant digits of the remaining term X_{m-1} at any m-th stage.

It is obvious that a similar method can be used to compute the square root in number systems other than the decimal number system. For instance, finding the digit-by-digit square root in the binary number system is quite efficient since the value of a_i is searched from a smaller set of binary digits {0,1}. This makes the computation faster since at each stage the value of Y_m is either $Y_m = 0$ for $a_m = 0$ or $Y_m = 2P_{m-1} + 1$ for $a_m = 1$. The fact that we have only two possible options for a_m also makes the process of deciding the value of a_m at m-th stage of calculation easier. This is because we only need to check if $Y_m \le X_{m-1}$ for $a_m = 1$. If this condition is satisfied, then we take $a_m = 1$; if not then $a_m = 0$. Also, the fact that multiplication by 2 is done by left bit-shifts helps in the computation.

Decimal (Base 10)

Write the original number in decimal form. The numbers are written similar to the long division algorithm, and, as in long division, the root will be written on the line above. Now separate the digits into pairs, starting from the decimal point and going both left and right. The decimal point of the root will be above the decimal point of the square. One digit of the root will appear above each pair of digits of the square.

Beginning with the left-most pair of digits, do the following procedure for each pair:

- Starting on the left, bring down the most significant (leftmost) pair of digits not yet used (if all the digits have been used, write "00") and write them to the right of the remainder from the previous step (on the first step, there will be no remainder). In other words, multiply the remainder by 100 and add the two digits. This will be the current value c.
- Find *p*, *y* and *x*, as follows:
    - Let *p* be the part of the root found so far, ignoring any decimal point. For the first step, *p* = 0.
    - Determine the greatest digit *x* such that $x(20p + x) \le c$. We will use a new variable $y = x(20p + x)$.
        - $20p + x$ is simply twice *p*, with the digit *x* appended to the right).
        - *x* can be found by guessing what $c/(20 \cdot p)$ is and doing a trial calculation of *y*, then adjusting *x* upward or downward as necessary.
    - Place the digit *x* as the next digit of the root, i.e., above the two digits of the square you just brought down. Thus the next *p* will be the old *p* times 10 plus *x*.

- Subtract y from c to form a new remainder.
- If the remainder is zero and there are no more digits to bring down, then the algorithm has terminated. Otherwise go back to step 1 for another iteration.

Examples: Find the Square Root of 152.2756.

```
          1  2. 3  4
        /-----------
     \/  01 52.27 56

       01                   1*1 <= 1 < 2*2                x = 1
       01                     y = x*x = 1*1 = 1
       --------
       00 52                22*2 <= 52 < 23*3             x = 2
       00 44                  y = (20+x)*x = 22*2 = 44
       --------
          08 27             243*3 <= 827 < 244*4          x = 3
          07 29               y = (240+x)*x = 243*3 = 729
          --------
             98 56          2464*4 <= 9856 < 2465*5       x = 4
             98 56            y = (2460+x)*x = 2464*4 = 9856
             --------
             00 00          Algorithm terminates: Answer is 12.34
```

Binary Numeral System (Base 2)

Inherent to digit-by-digit algorithms is a search and test step: find a digit, e, when added to the right of a current solution r, such that $(r+e)\cdot(r+e)\leq x$, where x is the value for which a root is desired. Expanding: $r\cdot r+2re+e\cdot e\leq x$. The current value of $r\cdot r$ —or, usually, the remainder—can be incrementally updated efficiently when working in binary, as the value of e will have a single bit set (a power of 2), and the operations needed to compute $2\cdot r\cdot e$ and $e\cdot e$ can be replaced with faster bit shift operations.

Example:

Here we obtain the square root of 81, which when converted into binary gives 1010001. The numbers in the left column gives the option between that number or zero to be used for subtraction at that stage of computation. The final answer is 1001, which in decimal is 9.

```
            1 0 0 1
           ---------
          √ 1010001

    1       1
            1
           ---------
    101      01
              0
            --------
    1001      100
                0
            --------
    10001     10001
              10001
             -------
                  0
```

This gives rise to simple computer implementations:

```
short isqrt(short num) {
    short res = 0;
    short bit = 1 << 14; // The second-to-top bit is set: 1 << 30 for
32 bits

    // "bit" starts at the highest power of four <= the argument.
    while (bit > num)
        bit >>= 2;

    while (bit != 0) {
        if (num >= res + bit) {
            num -= res + bit;
```

```
            res = (res >> 1) + bit;
        }
        else
            res >>= 1;
        bit >>= 2;
    }
    return res;
}
```

Using the notation above, the variable "bit" corresponds to e_m^2 which is $(2^m)^2 = 4^m$, the variable "res" is equal to $2re_m$, and the variable "num" is equal to the current X_m which is the difference of the number we want the square root of and the square of our current approximation with all bits set up to 2^{m+1}. Thus in the first loop, we want to find the highest power of 4 in "bit" to find the highest power of 2 in e. In the second loop, if num is greater than res + bit, then X_m is greater than $2re_m + e_m^2$ and we can subtract it. The next line, we want to add e_m to r which means we want to add $2e_m^2$ to $2re_m$ so we want res = res + bit<<1. Then update e_m to e_{m-1} inside res which involves dividing by 2 or another shift to the right. Combining these 2 into one line leads to res = res>>1 + bit. If X_m isn't greater than $2re_m + e_m^2$ then we just update e_m to e_{m-1} inside res and divide it by 2. Then we update e_m to e_{m-1} in bit by dividing it by 4. The final iteration of the 2nd loop has bit equal to 1 and will cause update of e to run one extra time removing the factor of 2 from res making it our integer approximation of the root.

Faster algorithms, in binary and decimal or any other base, can be realized by using lookup tables—in effect trading more storage space for reduced run time.

Exponential Identity

Pocket calculators typically implement good routines to compute the exponential function and the natural logarithm, and then compute the square root of S using the identity found using the properties of logarithms $(\ln x^n = n \ln x)$ and exponentials $(e^{\ln x} = x)$:

$$\sqrt{S} = e^{\frac{1}{2}\ln S}.$$

The denominator in the fraction corresponds to the nth root. In the case above the denominator is 2, hence the equation specifies that the square root is to be found. The same identity is used when computing square roots with logarithm tables or slide rules.

Two-variable Iterative Method

This method is applicable for finding the square root of $0 < \quad < 3$ and converges best for $S \approx 1$. This, however, is no real limitation for a computer based calculation, as in base 2 floating point and fixed point representations, it is trivial to multiply S by an integer power of 4, and therefore $\sqrt{S}$ by the corresponding power of 2, by changing the exponent or by shifting, respectively. Therefore, S can be moved to the range $\frac{1}{2} \leq S < 2$. Moreover, the following method does not employ general divisions, but only additions, subtractions, multiplications, and divisions by powers of two, which are again trivial to implement. A disadvantage of the method is that numerical errors accumulate, in contrast to single variable iterative methods such as the Babylonian one.

The initialization step of this method is:

$$a_0 = S$$
$$c_0 = S - 1$$

while the iterative steps read:

$$a_{n+1} = a_n - a_n c_n / 2$$
$$c_{n+1} = c_n^2 (c_n - 3) / 4$$

Then, $a_n \to \sqrt{S}$ (while $c_n \to 0$).

Note that the convergence of c_n, and therefore also of a_n, is quadratic.

The proof of the method is rather easy. First, rewrite the iterative definition of c_n as:

$$1 + c_{n+1} = (1 + c_n)(1 - c_n / 2)^2.$$

Then it is straightforward to prove by induction that:

$$S(1 + c_n) = a_n^2$$

and therefore the convergence of a_n to the desired result $\sqrt{S}$ is ensured by the convergence of c_n to 0, which in turn follows from $-1 < c_0 < 2$..

This method was developed around 1950 by M. V. Wilkes, D. J. Wheeler and S. Gill for use on EDSAC, one of the first electronic computers. The method was later generalized, allowing the computation of non-square roots.

Iterative Methods for Reciprocal Square Roots

The following are iterative methods for finding the reciprocal square root of S which is $1/\sqrt{S}$. Once it has been found, find $\sqrt{S}$ by simple multiplication: $\sqrt{S} = S\cdot(1/\sqrt{S})$. These iterations involve only multiplication, and not division. They are therefore faster than the Babylonian method. However, they are not stable. If the initial value is not close to the reciprocal square root, the iterations will diverge away from it rather than converge to it. It can therefore be advantageous to perform an iteration of the Babylonian method on a rough estimate before starting to apply these methods.

- Applying Newton's method to the equation $(1/x^2) - S = 0$ produces a method that converges quadratically using three multiplications per step:

$$x_{n+1} = \frac{x_n}{2}\cdot(3 - S\cdot x_n^2).$$

- Another iteration is obtained by Halley's method, which is the Householder's method of order two. This converges cubically, but involves four multiplications per iteration:

$$y_n = S\cdot x_n^2, \text{ and}$$

$$x_{n+1} = \frac{x_n}{8}\cdot(15 - y_n\cdot(10 - 3\cdot y_n)).$$

Goldschmidt's Algorithm

Some computers use Goldschmidt's algorithm to simultaneously calculate $\sqrt{S}$ and $1/\sqrt{S}$. Goldschmidt's algorithm finds $\sqrt{S}$ faster than Newton-Raphson iteration on a computer with a fused multiply–add instruction and either a pipelined floating point unit or two independent floating-point units.

The first way of writing Goldschmidt's algorithm begins:

$$b_0 = S$$

$$Y_0 \approx 1/\sqrt{S} \text{ (typically using a table lookup)}$$

$$y_0 = Y_0$$

$$x_0 = Sy_0$$

and iterates:

$$b_{n+1} = b_n Y_n^2$$

$$Y_{n+1} = (3 - b_{n+1})/2$$

$$x_{n+1} = x_n Y_{n+1}$$

$$y_{n+1} = y_n Y_{n+1}$$

until b_i is sufficiently close to 1, or a fixed number of iterations. The iterations converge to:

$$\lim_{n\to\infty} x_n = \sqrt{S}, \text{ and}$$
$$\lim_{n\to\infty} y_n = 1/\sqrt{S}.$$

Note that it is possible to omit either x_n and y_n from the computation, and if both are desired then $x_n = Sy_n$ may be used at the end rather than computing it through in each iteration.

A second form, using fused multiply-add operations, begins:

$$y_0 \approx 1/\sqrt{S} \text{ (typically using a table lookup)}$$
$$x_0 = Sy_0$$
$$h_0 = y_0/2$$

and iterates:

$$r_n = 0.5 - x_n h_n$$
$$x_{n+1} = x_n + x_n r_n$$
$$h_{n+1} = h_n + h_n r_n$$

until r_i is sufficiently close to 0, or a fixed number of iterations. This converges to:

$$\lim_{n\to\infty} x_n = \sqrt{S}, \text{ and}$$
$$\lim_{n\to\infty} 2h_n = 1/\sqrt{S}.$$

Taylor Series

If N is an approximation to $\sqrt{S}$, a better approximation can be found by using the Taylor series of the square root function:

$$\sqrt{N^2+d} = N\sum_{n=0}^{\infty}\frac{(-1)^n(2n)!}{(1-2n)n!^2 4^n}\frac{d^n}{N^{2n}} = N(1+\frac{d}{2N^2}-\frac{d^2}{8N^4}+\frac{d^3}{16N^6}-\frac{5d^4}{128N^8}+\cdots)$$

As an iterative method, the order of convergence is equal to the number of terms used. With two terms, it is identical to the Babylonian method. With three terms, each iteration takes almost as many operations as the Bakhshali approximation, but converges more slowly. Therefore, this is not a particularly efficient way of calculation. To maximize the rate of convergence, choose N so that $\frac{|d|}{N^2}$ is as small as possible.

Continued Fraction Expansion

Quadratic irrationals (numbers of the form $\frac{a+\sqrt{b}}{c}$, where a, b and c are integers), and in particular, square roots of integers, have periodic continued fractions. Sometimes what is desired is finding not the numerical value of a square root, but rather its continued fraction expansion, and hence its rational approximation. Let S be the positive number for which we are required to find the square root. Then assuming a to be a number that serves as an initial guess and r to be the remainder term, we can write $S = a^2 + r$. Since we have $S - a^2 = (\sqrt{S}+a)(\sqrt{S}-a) = r$, we can express the square root of S as:

$$\sqrt{S} = a + \frac{r}{a+\sqrt{S}}.$$

By applying this expression for $\sqrt{S}$ to the denominator term of the fraction, we have:

$$\sqrt{S} = a + \frac{r}{a + (a + \frac{r}{a+\sqrt{S}})} = a + \frac{r}{2a + \frac{r}{a+\sqrt{S}}}.$$

Proceeding this way, we get a generalized continued fraction for the square root as:

$$\sqrt{S} = a + \cfrac{r}{2a + \cfrac{r}{2a + \cfrac{r}{2a + \ddots}}}$$

The first step to evaluating such a fraction to obtain a root is to do numerical substitutions for the root of the number desired, and number of denominators selected. For example, in canonical form, r is 1 and for $\sqrt{2}$, a is 1, so the numerical continued fraction for 3 denominators is:

$$\sqrt{2} \approx 1 + \cfrac{1}{2 + \cfrac{1}{2 + \cfrac{1}{2}}}$$

Step 2 is to reduce the continued fraction from the bottom up, one denominator at a time, to yield a rational fraction whose numerator and denominator are integers. The reduction proceeds thus (taking the first three denominators):

$$1 + \cfrac{1}{2 + \cfrac{1}{2 + \cfrac{1}{2}}} = 1 + \cfrac{1}{2 + \cfrac{1}{\frac{5}{2}}}$$

$$=1+\cfrac{1}{2+\cfrac{2}{5}}=1+\cfrac{1}{\cfrac{12}{5}}$$

$$=1+\frac{5}{12}=\frac{17}{12}$$

Finally (step 3), divide the numerator by the denominator of the rational fraction to obtain the approximate value of the root:

$17 \div 12 = 1.42$ rounded to three digits of precision.

The actual value of $\sqrt{2}$ is 1.41 to three significant digits. The relative error is 0.17%, so the rational fraction is good to almost three digits of precision. Taking more denominators gives successively better approximations: four denominators yields the fraction $\frac{41}{29} = 1.4137$, good to almost 4 digits of precision, etc.

Usually, the continued fraction for a given square root is looked up rather than expanded in place because it's tedious to expand it. Continued fractions are available for at least square roots of small integers and common constants. For an arbitrary decimal number, precomputed sources are likely to be useless. The following is a table of small rational fractions called *convergents* reduced from canonical continued fractions for the square roots of a few common constants:

$\sqrt{S}$	Cont. fraction	~decimal	Serial convergents
$\sqrt{2}$	$[1;\overline{2}]$	1.41421	$\frac{3}{2},\frac{7}{5},\frac{17}{12},\frac{41}{29},\frac{99}{70},\frac{140}{99}$
$\sqrt{3}$	$[1;\overline{1,2}]$	1.73205	$\frac{2}{1},\frac{5}{3},\frac{7}{4},\frac{19}{11},\frac{26}{15},\frac{71}{41},\frac{97}{56}$
$\sqrt{5}$	$[2;\overline{4}]$	2.23607	$\frac{9}{4},\frac{38}{17},\frac{161}{72}$
$\sqrt{6}$	$[2;\overline{2,4}]$	2.44949	$\frac{5}{2},\frac{22}{9},\frac{49}{20},\frac{218}{89}$
$\sqrt{10}$	$[3;\overline{6}]$	3.16228	$\frac{19}{6},\frac{117}{37}$
$\sqrt{\pi}$	$[1;1,3,2,1,1,6...]$	1.77245	$\frac{2}{1},\frac{7}{4},\frac{16}{9},\frac{23}{13},\frac{39}{22}$

$\sqrt{e}$	[1;1,1,1,5,1,1...]	1.64872	$\frac{2}{1},\frac{3}{2},\frac{8}{5},\frac{28}{17},\frac{33}{20},\frac{61}{37}$
$\sqrt{\phi}$	[1;3,1,2,11,3,7...]	1.27202	$\frac{4}{3},\frac{5}{4},\frac{14}{11}$

All convergents up to and including denominator 99 listed.

In general, the larger the denominator of a rational fraction, the better the approximation. It can also be shown that truncating a continued fraction yields a rational fraction that is the best approximation to the root of any fraction with denominator less than or equal to the denominator of that fraction - e.g., no fraction with a denominator less than or equal to 99 is as good an approximation to $\sqrt{2}$ as 140/99.

Lucas Sequence Method

The Lucas sequence of the first kind $U_n(P,Q)$ is defined by the recurrence relations:

$$U_n(P,Q) = \begin{cases} 0 & if\, n=0 \\ 1 & if\, n=1 \\ P \cdot U_{n-1}(P,Q) - Q \cdot U_{n-2}(P,Q) & Otherwise \end{cases}$$

and the characteristic equation of it is:

$$x^2 - P \cdot x + Q = 0,$$

it has the discriminant $D = P^2 - 4Q$ and the roots:

$$x_1 = \frac{P+\sqrt{D}}{2}, \quad x_2 = \frac{P-\sqrt{D}}{2},$$

all that yield the following positive value:

$$\lim_{n\to\infty} \frac{U_{n+1}}{U_n} = x_1,$$

so when we want $\sqrt{a}$, we can choose $P=2$ and $Q=1-a$, and then calculate $x_1 = 1+\sqrt{a}$ using U_{n+1} and U_n for large value of n. The most effective way to calculate U_{n+1} and U_n is:

$$\begin{bmatrix} U_n \\ U_{n+1} \end{bmatrix} = \begin{bmatrix} 0 & 1 \\ -Q & P \end{bmatrix} \cdot \begin{bmatrix} U_{n-1} \\ U_n \end{bmatrix} = \begin{bmatrix} 0 & 1 \\ -Q & P \end{bmatrix}^n \cdot \begin{bmatrix} U_0 \\ U_1 \end{bmatrix}$$

$$\begin{bmatrix} 0 & 1 \\ a-1 & 2 \end{bmatrix}^n \cdot \begin{bmatrix} 0 \\ 1 \end{bmatrix} = \begin{bmatrix} U_n \\ U_{n+1} \end{bmatrix}$$

then when $n \to \infty$:

$$\sqrt{a} = \frac{U_n + 1}{U_n} - 1$$

Approximations that Depend on the Floating Point Representation

A number is represented in a floating point format as $m \times b^p$ which is also called scientific notation. Its square root is $\sqrt{m} \times b^{p/2}$ and similar formulae would apply for cube roots and logarithms. On the face of it, this is no improvement in simplicity, but suppose that only an approximation is required: then just $b^{p/2}$ is good to an order of magnitude. Next, recognise that some powers, p, will be odd, thus for 3141.59 = 3.14159 × 10^3 rather than deal with fractional powers of the base, multiply the mantissa by the base and subtract one from the power to make it even. The adjusted representation will become the equivalent of 31.4159 × 10^2 so that the square root will be $\sqrt{31.4159} \times 10$.

A table with only three entries could be enlarged by incorporating additional bits of the mantissa. However, with computers, rather than calculate an interpolation into a table, it is often better to find some simpler calculation giving equivalent results. Everything now depends on the exact details of the format of the representation, plus what operations are available to access and manipulate the parts of the number. For example, Fortran offers an EXPONENT(x) function to obtain the power. Effort expended in devising a good initial approximation is to be recouped by thereby avoiding the additional iterations of the refinement process that would have been needed for a poor approximation. Since these are few (one iteration requires a divide, an add, and a halving) the constraint is severe.

Many computers follow the IEEE (or sufficiently similar) representation, and a very rapid approximation to the square root can be obtained for starting Newton's method. The technique that follows is based on the fact that the floating point format (in base two) approximates the base-2 logarithm. That is $\log_2(m \times 2^p) = p + \log_2(m)$.

So for a 32-bit single precision floating point number in IEEE format (where notably, the power has a bias of 127 added for the represented form) you can get the approximate logarithm by interpreting its binary representation as a 32-bit integer, scaling it by 2^{-23}, and removing a bias of 127, i.e.

$$x_{int} \cdot 2^{-23} - 127 \approx \log_2(x).$$

For example, 1.0 is represented by a hexadecimal number 0x3F800000, which would represent $1065353216 = 127 \cdot 2^{23}$ if taken as an integer. Using the formula above you get $1065353216 \cdot 2^{-23} - 127 = 0$,, as expected from $\log_2(1.0)$. In a similar fashion you get 0.5 from 1.5 (0x3FC00000).

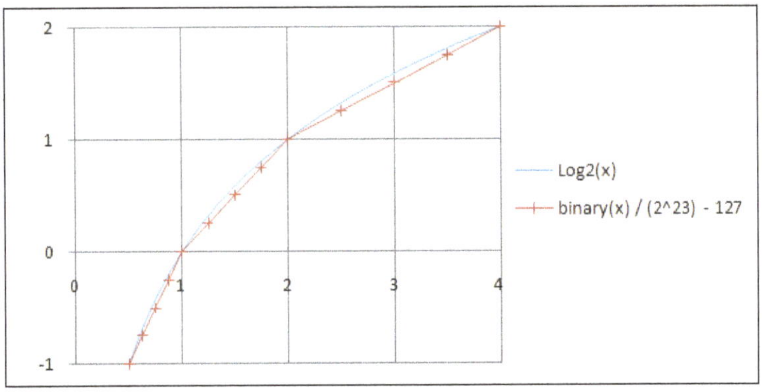

To get the square root, divide the logarithm by 2 and convert the value back. The following program demonstrates the idea. Note that the exponent's lowest bit is intentionally allowed to propagate into the mantissa. One way to justify the steps in this program is to assume b is the exponent bias and n is the number of explicitly stored bits in the mantissa and then show that:

$$(((x_{int} / 2^n - b) / 2) + b) \cdot 2^n = (x_{int} - 2^n) / 2 + ((b + 1) / 2) \cdot 2^n.$$

```
/* Assumes that float is in the IEEE 754 single precision floating point
format

 * and that int is 32 bits. */

float sqrt_approx(float z) {

    int val_int = *(int*)&z; /* Same bits, but as an int */

    /*

     * To justify the following code, prove that

     *

     * ((((val_int / 2^m) - b) / 2) + b) * 2^m = ((val_int - 2^m) / 2)
+ ((b + 1) / 2) * 2^m)

     *

     * where

     *

     * b = exponent bias

     * m = number of mantissa bits

     *
```

```
     * .
     */

    val_int -= 1 << 23; /* Subtract 2^m. */
    val_int >>= 1; /* Divide by 2. */
    val_int += 1 << 29; /* Add ((b + 1) / 2) * 2^m. */

    return *(float*)&val_int; /* Interpret again as float */
}
```

The three mathematical operations forming the core of the above function can be expressed in a single line. An additional adjustment can be added to reduce the maximum relative error. So, the three operations, not including the cast, can be rewritten as:

```
val_int = (1 << 29) + (val_int >> 1) - (1 << 22) + a;
```

where a is a bias for adjusting the approximation errors. For example, with a = 0 the results are accurate for even powers of 2 (e.g., 1.0), but for other numbers the results will be slightly too big (e.g.,1.5 for 2.0 instead of 1.414... with 6% error). With a = -0x4B0D2, the maximum relative error is minimized to ±3.5%.

If the approximation is to be used for an initial guess for Newton's method to the equation $(1/x^2)-S=0$, then the reciprocal form shown in the following section is preferred.

Reciprocal of the Square Root

A variant of the above routine is included below, which can be used to compute the reciprocal of the square root, i.e., $x^{-\frac{1}{2}}$ instead, was written by Greg Walsh. The integer-shift approximation produced a relative error of less than 4%, and the error dropped further to 0.15% with one iteration of Newton's method on the following line. In computer graphics it is a very efficient way to normalize a vector.

```
float invSqrt(float x) {
    float xhalf = 0.5f*x;
    union {
        float x;
        int i;
    } u;
    u.x = x;
```

```
    u.i = 0x5f375a86 - (u.i >> 1);

    /* The next line can be repeated any number of times to increase
accuracy */

    u.x = u.x * (1.5f - xhalf * u.x * u.x);

    return u.x;

}
```

Some VLSI hardware implements inverse square root using a second degree polynomial estimation followed by a Goldschmidt iteration.

Negative or Complex Square

If $S < 0$, then its principal square root is:

$$\sqrt{S} = \sqrt{|S|}\,i.$$

If $S = a+bi$ where a and b are real and $b \neq 0$, then its principal square root is:

$$\sqrt{S} = \sqrt{\frac{|S|+a}{2}} + \operatorname{sgn}(b)\sqrt{\frac{|S|-a}{2}}\,i.$$

This can be verified by squaring the root. Here:

$$|S| = \sqrt{a^2+b^2}$$

is the modulus of S. The principal square root of a complex number is defined to be the root with the non-negative real part.

ROOT-FINDING ALGORITHM

Fixed-point Iteration

In numerical analysis, fixed-point iteration is a method of computing fixed points of iterated functions.

More specifically, given a function f defined on the real numbers with real values and given a point x_0 in the domain of f, the fixed point iteration is

$$x_{n+1} = f(x_n), n = 0,1,2,\ldots$$

which gives rise to the sequence $x_0, x_1, x_2, \ldots$ which is hoped to converge to a point x. If f is continuous, then one can prove that the obtained x is a fixed point of f, i.e.,

$$f(x) = x.$$

More generally, the function f can be defined on any metric space with values in that same space.

Examples:

- A first simple and useful example is the Babylonian method for computing the square root of $a>0$, which consists in taking $f(x)=\frac{1}{2}\left(\frac{a}{x}+x\right)$, i.e. the mean value of x and a/x, to approach the limit $x=\sqrt{a}$ (from whatever starting point $x_0 \gg 0$). This is a special case of Newton's method.

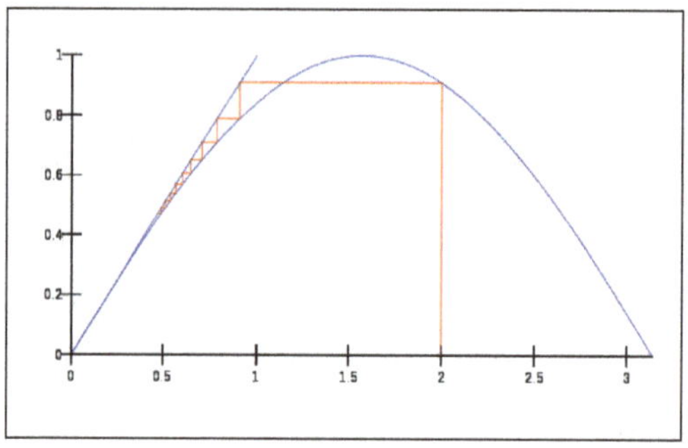

The fixed-point iteration $x_{n+1} = \sin x_n$ with initial value $x_0 = 2$ converges to 0. This example does not satisfy the assumptions of the Banach fixed point theorem and so its speed of convergence is very slow.

- The fixed-point iteration $x_{n+1} = \cos x_n$ converges to the unique fixed point of the function $f(x) = \cos x$ for any starting point x_0. This example does satisfy the assumptions of the Banach fixed point theorem. Hence, the error after n steps satisfies $|x_n - x| \le \frac{q^n}{1-q}|x_1 - x_0| = Cq^n$ (where we can take $q = 0.85$, if we start from $x_0 = 1$.) When the error is less than a multiple of q^n for some constant q, we say that we have linear convergence. The Banach fixed-point theorem allows one to obtain fixed-point iterations with linear convergence.
- The fixed-point iteration $x_{n+1} = 2x_n$ will diverge unless $x_0 = 0$. We say that the fixed point of $f(x) = 2x$ is repelling.
- The requirement that f is continuous is important, as the following example shows. The iteration:

$$x_{n+1} = \begin{cases} \frac{x_n}{2}, & x_n \neq 0 \\ 1, & x_n = 0 \end{cases}$$

converges to 0 for all values of x_0. However, 0 is *not* a fixed point of the function:

$$f(x)=\begin{cases}\frac{x}{2}, & x\neq 0\\ 1, & x=0\end{cases}$$

as this function is *not* continuous at $x=0$, and in fact has no fixed points.

Applications

- Newton's method for finding roots of a given differentiable function $f(x)$ is:

  $$x_{n+1}=x_n-\frac{f(x_n)}{f'(x_n)}.$$

  If we write $g(x)=x-\frac{f(x)}{f'(x)}$, we may rewrite the Newton iteration as the fixed-point iteration

  $$x_{n+1}=g(x_n).$$

  If this iteration converges to a fixed point *x* of *g*, then:

  $$x=g(x)=x-\frac{f(x)}{f'(x)},\quad \text{so } f(x)/f'(x)=0.$$

  The reciprocal of anything is nonzero, therefore *f*(*x*) = 0: *x* is a *root* of *f*. Under the assumptions of the Banach fixed point theorem, the Newton iteration, framed as the fixed point method, demonstrates linear convergence. However, a more detailed analysis shows quadratic convergence, i.e.,

  $|x_n-x|<Cq^{2^n}$, under certain circumstances.

- Halley's method is similar to Newton's method but, when it works correctly, its error is $|x_n-x|<Cq^{3^n}$ (cubic convergence). In general, it is possible to design methods that converge with speed Cq^{k^n} for any $k\in\mathbb{N}$. As a general rule, the higher the *k*, the less stable it is, and the more computationally expensive it gets. For these reasons, higher order methods are typically not used.

- Runge–Kutta methods and numerical ordinary differential equation solvers in general can be viewed as fixed point iterations. Indeed, the core idea when analyzing the A-stability of ODE solvers is to start with the special case $y'=ay$, where a is a complex number, and to check whether the ODE solver converges to the fixed point $y=0$ whenever the real part of a is negative.

- The Picard–Lindelöf theorem, which shows that ordinary differential equations have solutions, is essentially an application of the Banach fixed point theorem to a special sequence of functions which forms a fixed point iteration, constructing the solution to the equation. Solving an ODE in this way is called Picard iteration, Picard's method, or the Picard iterative process.

- The iteration capability in Excel can be used to find solutions to the Colebrook equation to an accuracy of 15 significant figures.
- Some of the "successive approximation" schemes used in dynamic programming to solve Bellman's functional equation are based on fixed point iterations in the space of the return function.

Contractions

If a function f defined on the real line with real values is Lipschitz continuous with Lipschitz constant $L < 1$, then this function has precisely one fixed point, and the fixed-point iteration converges towards that fixed point for any initial guess x_0. This theorem can be generalized to any metric space.

Proof of this theorem:

Since f is Lipschitz continuous with Lipschitz constant $L < 1$, then for the sequence $\{x_n, n = 0,1,2,\ldots\}$, we have:

$$\begin{aligned} |x_2 - x_1| &= |f(x_1) - f(x_0)| \le L|x_1 - x_0|, \\ |x_3 - x_2| &= |f(x_2) - f(x_1)| \le L|x_2 - x_1|, \\ &\vdots \end{aligned}$$

and,

$$|x_n - x_{n-1}| = |f(x_{n-1}) - f(x_{n-2})| \le L|x_{n-1} - x_{n-2}|.$$

Combining the above inequalities yields:

$$|x_n - x_{n-1}| \le L^{n-1}|x_1 - x_0|.$$

Since $L < 1$, $L^{n-1} \to 0$ as $n \to \infty$.

Therefore, we can show $\{x_n\}$ is a Cauchy sequence and thus it converges to a point x^*.

For the iteration $x_n = f(x_{n-1})$, let n go to infinity on both sides of the equation, we obtain $x^* = f(x^*)$. This shows that x^* is the fixed point for f. So we proved the iteration will eventually converge to a fixed-point.

This property is very useful because not all iterations can arrive at a convergent fixed-point. When constructing a fixed-point iteration, it is very important to make sure it converges. There are several fixed-point theorems to guarantee the existence of the fixed point, but since the iteration function is continuous, we can usually use the above theorem to test if an iteration converges or not. The proof of the generalized theorem to metric space is similar.

Convergence Acceleration

The speed of convergence of the iteration sequence can be increased by using a convergence acceleration method such as Aitken's delta-squared process. The application of Aitken's method to fixed-point iteration is known as Steffensen's method, and it can be shown that Steffensen's method yields a rate of convergence that is at least quadratic.

Real-root Isolation

In numerical analysis and computer algebra, real-root isolation of a polynomial consist of producing disjoint intervals of the real line, which contain each one (and only one) real root of the polynomial, and, together, contain all the real roots of the polynomial.

Real-root isolation is useful because usual root-finding algorithms for computing the real roots of a polynomial may produce some real roots, but, cannot generally certify having found all real roots. In particular, if such an algorithm does not find any root, one does not know whether it is because there is no real root. Some algorithms compute all complex roots, but, as there are generally much fewer real roots than complex roots, most of their computation time is generally spent for computing non-real roots (in the average, a polynomial of degree n has n complex roots, and only $\log n$ real roots). Moreover, it may be difficult to distinguish the real roots from the non-real roots with small imaginary part.

The first complete real-root isolation algorithm results from Sturm's theorem. However, when real-root-isolation algorithms began to be implemented on computers it appeared that algorithms derived from Sturm's theorem are less efficient than those derived from Descartes' rule of signs.

Since the beginning of 20th century there is an active research activity for improving the algorithms derived from Descartes' rule of signs, getting very efficient implementations, and computing their computational complexity. The best implementations can routinely isolate real roots of polynomials of degree more than 1,000.

Specifications and General Strategy

For finding real roots of a polynomial, the common strategy is to divide the real line (or an interval of it where root are searched) into disjoint intervals until having at most one root in each interval. Such a procedure is called root isolation, and a resulting interval that contains exactly one root is an isolating interval for this root.

Wilkinson's polynomial shows that a very small modification of one coefficient of a polynomial may change dramatically not only the value of the roots, but also their nature (real or complex). Also, even with a good approximation, when one evaluates a polynomial at an approximate root, one may get a result that is far to be close to zero. For example, if a polynomial of degree 20 (the degree of Wilkinson's polynomial) has a root close to 10, the derivative of the polynomial at the root may be of the order of 10^{20}; this implies that an error of 10^{-10} on the value of the root may produce a value of the polynomial at the approximate root that is of the order of 10^{10}. It follows that, except maybe for very low degrees, a root-isolation procedure cannot give reliable results without using exact arithmetic. Therefore, if one wants to isolate roots of a polynomial with floating-point coefficients, it is often better to convert them to rational numbers, and then take the primitive part of the resulting polynomial, for having a polynomial with integer coefficients.

For this reason, although the methods that are described later work theoretically with real numbers, they are generally used in practice with polynomials with integer coefficients, and intervals ending with rational numbers. Also, the polynomials are always supposed to be square free. There are two reasons for that. Firstly Yun's algorithm for computing the square-free factorization is less costly than twice the cost of the computation of the greatest common divisor of the polynomial and its derivative. As this may produce factors of lower degrees, it is generally advantageous to apply root-isolation algorithms only on polynomials without multiple roots, even when this is not required by the algorithm. The second reason for considering only square-free polynomials is that the fastest root-isolation algorithms do not work in the case of multiple roots.

For root isolation, one requires a procedure for counting the real roots of a polynomial in an interval without having to compute them, or, at least a procedure for deciding whether an interval contains zero, one or more roots. With such a decision procedure, one may work with a working list of intervals that may contain real roots. At the beginning, the list contains a single interval containing all roots of interest, generally the whole real line or its positive part. Then each interval of the list is divided into two smaller intervals. If one of the new intervals does not contains any root, it is removed from the list. If it contains one root, it is put in an output list of isolating intervals. Otherwise, it is kept in the working list for further divisions, and the process may continue until all roots are eventually isolated.

Sturm's Theorem

The first complete root-isolation procedure results of Sturm's theorem, which expresses the number of real roots in an interval in terms of the number of sign variations of the values of a sequence of polynomials, called *Sturm's sequence*, at the ends of the interval. Sturm's sequence is the sequence of remainders that occur in a variant of Euclidean algorithm applied to the polynomial and its derivatives. When implemented on computers, it appeared that root isolation with Sturm's theorem is less efficient than the other methods that is described later. Consequently, Sturm's theorem is rarely used for effective computations, although it remains useful for theoretical purpose.

Descartes' Rule of Signs and its Generalizations

Descartes' rule of signs asserts that the difference between the number of sign variations in the sequence of the coefficients of a polynomial and the number of its positive real roots is a nonnegative even integer. It results that if this number of sign variations is zero, then the polynomial does not have any positive real roots, and, if this number is one, then the polynomial has a unique positive real root, which is a single root. Unfortunately the converse is not true, that is, a polynomial which has either no positive real root or as a single positive simple root may have a number of sign variations greater than 1.

This has been generalized by Budan's theorem, into a similar result for the real roots in a half-open interval $(a, b]$: If $f(x)$ is a polynomial, and v is the difference between of the numbers of sign variations of the sequences of the coefficients of $f(x + a)$ and $f(x + b)$, then v minus the number of real roots in the interval, counted with their multiplicities, is a nonnegative even integer. This is a generalization of Descartes' rule of signs, because, for b sufficiently large, there is no sign variation in the coefficients of $f(x + b)$, and all real roots are smaller than b.

Budan's may provide a real-root-isolation algorithm for a square-free polynomial (a polynomial without multiple root): from the coefficients of polynomial, one may compute an upper bound M of the absolute values of the roots and a lower bound m on the absolute values of the differences of two roots. Then, if one divides the interval $[-M, M]$ into intervals of length less than m, then every real root is contained in some interval, and no interval contains two roots. The isolating intervals are thus the intervals for which Budan's theorem asserts an odd number of roots.

However, this algorithm is very inefficient, as one cannot use a coarser partition of the interval $[-M, M]$, because, if Budan's theorem gives a result larger than 1 for an interval of larger size, there is no way for insuring that it does not contain several roots.

Vincent's and Related Theorems

Vincent's theorem provides a method for real-root isolation, which is at the basis of the most efficient real-root-isolation algorithms. It concerns the positive real roots of a square-free polynomial (that is a polynomial without multiple roots). If $a_1, a_2, \ldots,$ is a sequence of positive real numbers, let:

$$c_k = a_1 + \cfrac{1}{a_2 + \cfrac{1}{a_3 + \cfrac{1}{\ddots + \cfrac{1}{a_k}}}}$$

be the kth convergent of the continued fraction:

$$a_1 + \cfrac{1}{a_2 + \cfrac{1}{a_3 + \ddots}}.$$

Vincent's theorem — Let $p_0(x)$ be a square-free polynomial of degree n, and $a_1, a_2, \ldots,$ be a sequence of real numbers. For $i = 1, 2, \ldots,$ consider the polynomial:

$$p_i(x) = x^n p_{i-1}\left(a_i + 1/x\right).$$

Then, there is an integer k such that either $p_k(0) = 0$, or the sequence of the coefficients of p_k has at most one sign variation.

In the first case, the convergent c_k is a positive root of p_0. Otherwise, this number of sign variations (either 0 or 1) is the number of real roots of p_0 in the interval defined by c_{k-1} and c_k.

For proving his theorem, Vincent proved a result that is useful on its own:

Vincent's auxiliary theorem — If $p(x)$ is a square-free polynomial of degree n, and a, b, c, d are nonnegative real numbers such that $\left|\frac{a}{c} - \frac{b}{d}\right|$ is small enough (but not 0), then there is at most one sign variation in the coefficients of the polynomial:

$$q(x) = (cx + d)^n p\left(\frac{ax + b}{cx + d}\right),$$

and this number of sign variations is the number of real roots of $p(x)$ in the open interval defined by $\frac{a}{b}$ and $\frac{b}{d}$.

For working with real numbers, one may always choose $c = d = 1$, but, as effective computations are done with rational numbers, it is generally convenient to suppose that a, b, c, d are integers.

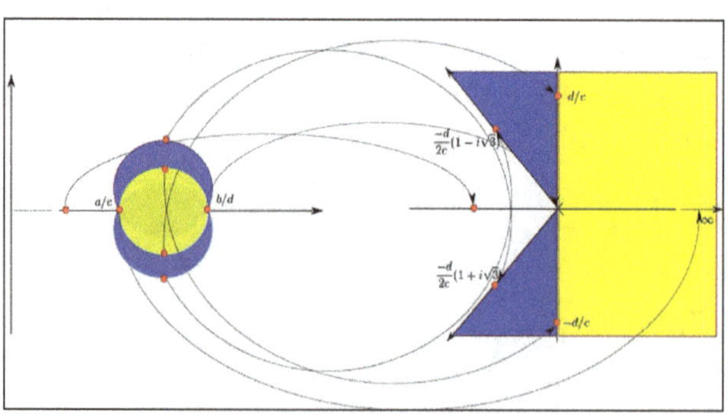

The "small enough" condition has been quantified independently by Nikola Obreshkov, and Alexander Ostrowski.

Obreschkoff–Ostrowski theorem: In blue and yellow, the regions of the complex plane where there should be no root for having 0 or 1 sign variation; on the left the regions excluded for the roots of p, on the right, the regions excluded for the roots of the transformed polynomial q; in blue, the regions that are excluded for having one sign variation, but allowed for zero sign variations.

Theorem (Obreschkoff–Ostrowski) — The conclusion of Vincent's auxiliary result holds if the polynomial $p(x)$ has at most one root $\alpha + i\beta$ such that:

$$\left(\alpha-\frac{a}{c}\right)\left(\alpha-\frac{b}{d}\right)+\beta^2 \le \frac{1}{\sqrt{3}}\left|\beta\left(\frac{a}{c}-\frac{b}{d}\right)\right|.$$

In particular the conclusion holds if:

$$\left|\frac{a}{c}-\frac{b}{d}\right| < \frac{\operatorname{sep}(p)}{2\sqrt{3}}$$

where sep(p) is the minimal distance between two roots of p.

For polynomials with integer coefficients, the minimum distance sep(p) may be lower bounded in terms of the degree of the polynomial and the maximal absolute value of its coefficients. This allows the analysis of worst-case complexity of algorithms based on Vincent's theorems. However, Obreschkoff–Ostrowski theorem shows that the number of iterations of these algorithms depend on the distances between roots in the neighborhood of the working interval; therefore, the number of iterations may vary dramatically for different roots of the same polynomial.

James V. Uspensky gave a bound on the length of the continued fraction (the integer k needed, in Vincent's theorem, for getting zero or one sign variations:

Theorem (Uspensky) — Let $p(x)$ be a polynomial of degree n, and sep(p) be the minimal distance between two roots of p. Let:

$$\varepsilon = \left(n+\frac{1}{n}\right)^{\frac{1}{n-1}} - 1.$$

Then the integer k, whose existence is asserted in Vincent's theorem, is not greater than the smallest integer h such that:

$$F_{h-1}\operatorname{sep}(p) > 2 \quad and \quad F_{h-1}F_k \operatorname{sep}(p) > \frac{1}{\mathring{a}},$$

where F_h is the hth Fibonacci number.

Continued Fraction Method

The use of continued fractions for real-root isolation has been introduced by Vincent, although he credited Joseph-Louis Lagrange for this idea, without providing a reference. For making an algorithm of Vincent's theorem, one must provide a criterion for choosing the a_i that occur in his theorem. Vincent himself provided some choice. Some other choices are possible, and the efficiency of the algorithm may depend dramatically on these choices. Below is presented an algorithm, in which these choices result from an auxiliary function.

For running this algorithm one must work with a list of intervals represented by a specific data structure. The algorithm works by choosing an interval, removing it from the list, adding zero, one or two smaller intervals to the list, and possibly outputs an isolation interval.

For isolating the real roots of a polynomial $p(x)$ of degree n, each interval is represented by a pair $(A(x), M(x))$, where $A(x)$ is a polynomial of degree n and $M(x) = \frac{px+r}{qx+s}$ is a Möbius transformation with integer coefficients. One has:

$$A(x) = p(M(x)),$$

and the interval represented by this data structure is the interval that has $M(\infty) = \frac{p}{q}$ and $M(0) = \frac{r}{s}$ as end points. The Möbius transformation maps the roots of p in this interval to the roots of A in $(0, +\infty)$.

The algorithm works with a list of intervals that, at the beginning, contains the two intervals $(A(x) = p(x), M(x) = x)$ and $(A(x) = p(-x), M(x) = -x)$, corresponding to the partition of the reals into the positive and the negative ones (one may suppose that zero is not a root, as, if it were, it suffices to apply the algorithm to $p(x)/x$). Then for each interval $(A(x), M(x))$ in the list, the algorithm remove it from the list; if the number of sign variations of the coefficients of A is zero, there is no root in the interval, and one passes to the next interval. If the number of sign variations is one, the interval defined by $M(0)$ and $M(\infty)$ is an isolating interval. Otherwise, one chooses a positive real number b for dividing the interval $(0, +\infty)$ into $(0, b)$ and $(b, +\infty)$, and, for each subinterval, one composes M with a Möbius transformation that maps the interval onto $(0, +\infty)$, for getting two new intervals to be added to the list. In pseudocode, this gives the following, where var(A) denotes the number of sign variations of the coefficients of the polynomial A.

```
function continued fraction is

   input: P(x), a square-free polynomial,

   output: a list of pairs of rational numbers defining isolating in-
tervals
```

```
   /* Initialization */
   L := [(P(x), x), (P(-x), -x)]                     /* two starting inter-
vals */
   Isol := [ ]
    /* Computation */
   while L ≠ [ ] do
      Choose (A(x), M(x)) in L, and remove it from L
      v := var(A)
      if v = 0 then exit                              /* no root in the in-
terval */
      if v = 1 then                                   /* isolating interval
found */
         add (M(0), M(∞)) to Isol
         exit
      b := some positive integer
      B(x) = A(x + b)
      w := v - var(B)
      if B(0) = 0 then                           /* rational root found */
         add (M(b), M(b)) to Isol
         B(x) := B(x)/x
      add (B(x),  M(b + x) to L                     /* roots in (b, +∞) */
      if w = 0 then exit                            /* Budan's theorem */
       if w = 1 then                                   /* Budan's theorem
again */
         add (M(0), M(b)) to Isol
      if w > 1 then
        add A(b/(1 + x)),  M(b/(1 + x)) to L       /* roots in (0, b) */
```

The different variants of the algorithm depend essentially on the choice of b.

The drawback of always choosing $b = 1$ is that one has to do many successive changes of variable of the form $x \to 1 + x$. These may be replaced by a single change of variable $x \to n + x$, but, nevertheless, one has to do the intermediate changes of variables for applying Budan's theorem.

A way for improving the efficiency of the algorithm is to take for b a lower bound of the positive real roots, computed from the coefficients of the polynomial.

Bisection Method

The bisection method consists roughly of starting from an interval containing all real roots of a polynomial, and divides it recursively into two parts until getting eventually intervals that contain either zero of one root. The starting interval may be of the form $(-B, B)$, where B is an upper bound on the absolute values of the roots, such as those that are given in Properties of polynomial roots § Bounds on (complex) polynomial roots. For technical reasons (simpler changes of variable, simpler complexity analysis, possibility of taking advantage of the binary analysis of computers), the algorithms are generally presented as starting with the interval [0, 1]. There is no loss of generality, as the changes of variables $x = By$ and $x = -By$ move respectively the positive and the negative roots in the interval [0, 1]. (The single changes variable $x = (2By - B)$ may also be used).

The method requires an algorithm for testing whether an interval has zero, one, or possibly several roots, and for warranting termination, this testing algorithm must excludes the possibility of getting infinitely many times the output "possibility of several roots". Sturm's theorem and Vincent's auxiliary theorem provide such convenient tests. As the use Descartes' rule of signs and Vincent's auxiliary theorem is much more computationally efficient than the use of Sturm's theorem.

The bisection method based on Descartes' rules of signs and Vincent's auxiliary theorem has been introduced in 1976 by Akritas and Collins under the name of Modified Uspensky algorithm, and has been referred to as Uspensky algorithm, Vincent–Akritas–Collins algorithm, or Descartes method, although Descartes, Vincent and Uspensky have never described it.

The method works as follows. For searching the roots in some interval, one changes first the variable for mapping the interval onto [0, 1] giving a new polynomial $q(x)$. For searching the roots of q in [0, 1], one maps the interval [0, 1] onto [0, +∞]) by the change of variable $x \to \frac{1}{x+1}$, giving a polynomial $r(x)$. Descartes' rule of signs applied to the polynomial r gives indications on the number of real roots of q in the interval [0, 1], and thus on the number of roots of the initial polynomial in the interval that has been mapped on [0, 1]. If there is no sign variation in the sequence of the coefficients of r, then there is no real root in the considered intervals. If there is one sign variation, then one has an isolation interval. Otherwise, one splits the interval [0, 1] into [0, 1/2] and [1/2, 1], one maps them onto [0, 1] by the changes of variable $x = y/2$ and $x = (y + 1)/2$. Vincent's auxiliary theorem insures the termination of this procedure.

Except for the initialization, all these changes of variable consists of the composition of at most two very simple changes of variable which are the scalings by two $x \to x/2$, the translation $x \to x + 1$, and the inversion $x \to 1/x$, the latter consisting simply of reverting the order of the coefficients of the polynomial. As most of the computing time is devoted to changes of variable, the method consisting of mapping eery interval to [0, 1] is fundamental for insuring a good efficiency.

Pseudocode

The following notation is used in the pseudocode that follows.

- $p(x)$ is the polynomial for which the real roots in the interval [0, 1] have to be isolated.
- var($q(x)$) denotes the number of sign variations in the sequence of the coefficients of the polynomial q.
- The elements of working list have the form $(c, k, q(x))$, where:
    - c and k are two nonnegative integers such that $c < 2^k$, which represent the interval $\left[\frac{c}{2^k}, \frac{c+1}{2^k}\right]$.
    - $q(x) = 2^{kn} p\left(\frac{x+c}{2^k}\right)$, where n is the degree of p (the polynomial q may be computed directly from p, c and k, but it is less costly to compute it incrementally, as it will be done in the algorithm; if p has integer coefficients, the same is true for q).

```
function bisection is

    input: p(x), a square-free polynomial, such that p(0) p(1) ≠ 0,

        for which the roots in the interval [0, 1] are searched

    output: a list of triples (c, k, h),

        representing isolating intervals of the form [c/2^k, (c+1)/2^k]

    /* Initialization */

    L := [(0, 0, p(x))] /* a single element in the working list L */

    Isol := [ ]

    n := degree(p}}

    /* Computation */
```

```
while L ≠ [ ] do
    Choose (c, k, q(x)) in L, and remove it from L
    if q(0) = 0 then
        q(x) := q(x)/x
        n := n - 1                    /* A rational root found */
        add (c, k, 0) to Isol
    v := var((x + 1)ⁿ q(1/x + 1)))
    if v = 1 then                    /* An isolating interval found */
        add (c, k, 1) to Isol
    if v > 1 then                    /* Bisecting */
        add (2c, k + 1,    to L
        add (2c + 1, k + 1,    to L
end
```

This procedure is essentially the one that has been described by Collins and Akritas. The running time depends mainly on the number of intervals that have to be considered, and on the changes of variables. There are ways for improving the efficiency, which have been an active subject of research since the publication of the algorithm, and mainly since the beginning of 20th century.

Recent Improvements

Various ways for improving Akritas–Collins bisection algorithm have been proposed. They include a method for avoiding storing a long list of polynomials without losing the simplicity of the changes of variables, the use of approximate arithmetic (floating point and interval arithmetic) when it allows getting the right value for the number of sign variations, the use of Newton's method when possible, the use of fast polynomial arithmetic, shortcuts for long chains of bisections in case of clusters of close roots, bisections in unequal parts for limiting instability problems in polynomial evaluation.

All these improvement lead to an algorithm for isolating all real roots of a polynomial with integer coefficients, which has the complexity (using soft O notation, $\tilde{O}$, for omitting logarithmic factors):

$$\tilde{O}(n^2(k+t)),$$

where n is the degree of the polynomial, k is the number of nonzero terms, t is the maximum of digits of the coefficients.

The implementation of this algorithm appears to be more efficient than any other implemented method for computing the real roots of a polynomial, even in the case of polynomials having very close roots (the case which was previously the most difficult for the bisection method).

Steffensen's Method

In numerical analysis, Steffensen's method is a root-finding technique similar to Newton's method, named after Johan Frederik Steffensen. Steffensen's method also achieves quadratic convergence, but without using derivatives as Newton's method does.

The simplest form of the formula for Steffensen's method occurs when it is used to find the zeros, or roots, of a function f; that is: to find the value $x_\star$ that satisfies $f(x_\star)=0$. Near the solution $x_\star$, the function f is supposed to approximately satisfy $-1<f'(x_\star)<0$; this condition makes f adequate as a correction-function for for finding its *own* solution, although it is not required to work efficiently. For some functions, Steffensen's method can work even if this condition is not met, but in such a case, the starting value x_0 must be *very* close to the actual solution $x_\star$, and convergence to the solution may be slow.

Given an adequate starting value x_0, a sequence of values $x_0, x_1, x_2, \ldots, x_n, \ldots$ can be generated using the formula below. When it works, each value in the sequence is much closer to the solution $x_\star$ than the prior value. The value x_n from the current step generates the value x_{n+1} for the next step, via this formula:

$$x_{n+1} = x_n - \frac{f(x_n)}{g(x_n)}$$

for n = 0, 1, 2, 3, ... , where the slope function $g(x_n)$ is a composite of the original function f given by the following formula:

$$g(x_n) = \frac{f(x_n + f(x_n))}{f(x_n)} - 1$$

or equivalently:

$$g(x_n) = \frac{f(x_n + h) - f(x_n)}{h}$$

where,

$$h = f(x_n).$$

The function g is the average value for the slope of the function f between the last sequence point $(x,y)=(x_n, f(x_n))$ and the auxiliary point $(x,y)=(x_n+h, f(x_n+h))$, with the step $h = f(x_n)$. It is also called the first-order divided difference of f between those two points.

It is only for the purpose of finding h for this auxiliary point that the value of the function f must be an adequate correction to get closer to its own solution, and for that reason fulfill the requirement that $-1 < f'(x_\star) < 0$. For all other parts of the calculation, Steffensen's method only requires the function f to be continuous and to actually have a nearby solution. Several modest modifications of the step h in the slope calculation g exist to accommodate functions f that do not quite meet the requirement.

Advantages and Drawbacks

The main advantage of Steffensen's method is that it has quadratic convergence like Newton's method – that is, both methods find roots to an equation f just as 'quickly'. In this case *quickly* means that for both methods, the number of correct digits in the answer doubles with each step. But the formula for Newton's method requires evaluation of the function's derivative f' as well as the function f, while Steffensen's method only requires f itself. This is important when the derivative is not easily or efficiently available.

The price for the quick convergence is the double function evaluation: Both $f(x_n)$ and $f(x_n + h)$ must be calculated, which might be time-consuming if f is a complicated function. For comparison, the secant method needs only one function evaluation per step. The secant method increases the number of correct digits by "only" a factor of roughly 1.6 per step, but one can do twice as many steps of the secant method within a given time. Since the secant method can carry out twice as many steps in the same time as Steffensen's method,[a] when both algorithms succeed, the secant method converges faster than Steffensen's method in actual practice: The secant method achieves a factor of about $(1.6)^2 \approx 2.6$ times as many digits for every two steps (two function evaluations), compared to Steffensen's factor of 2 for every one step (two function evaluations).

Similar to most other iterative root-finding algorithms, the crucial weakness in Steffensen's method is the choice of the starting value x_0. If the value of x_0 is not 'close enough' to the actual solution $x_\star$, the method may fail and the sequence of values $x_0, x_1, x_2, x_3, \ldots$ may either flip-flop between two extremes, or diverge to infinity.

Derivation using Aitken's Delta-squared Process

The version of Steffensen's method implemented in the MATLAB code can be found using the Aitken's delta-squared process for accelerating convergence of a sequence. To compare the following formulae to the formulae in the section above, notice that $x_n = p - p_n$. This method assumes starting with a linearly convergent sequence and increases the rate of convergence of that sequence. If the signs of p_n, p_{n+1}, p_{n+2} agree and p_n is 'sufficiently close' to the desired limit of the sequence p, we can assume the following:

$$\frac{p_{n+1}-p}{p_n-p} \approx \frac{p_{n+2}-p}{p_{n+1}-p}$$

then,

$$(p_{n+1}-p)^2 \approx (p_{n+2}-p)(p_n-p)$$

so,

$$p_{n+1}^2-2p_{n+1}p+p^2 \approx p_{n+2}p_n-(p_n+p_{n+2})p+p^2$$

and hence,

$$(p_{n+2}-2p_{n+1}+p_n)p \approx p_{n+2}p_n-p_{n+1}^2.$$

Solving for the desired limit of the sequence p gives:

$$\begin{aligned} p &\approx \frac{p_{n+2}p_n-p_{n+1}^2}{p_{n+2}-2p_{n+1}+p_n} \\ &= \frac{p_n^2+p_np_{n+2}+2p_np_{n+1}-2p_np_{n+1}-p_n^2-p_{n+1}^2}{p_{n+2}-2p_{n+1}+p_n} \\ &= \frac{(p_n^2+p_np_{n+2}-2p_np_{n+1})-(p_n^2-2p_np_{n+1}+p_{n+1}^2)}{p_{n+2}-2p_{n+1}+p_n} \\ &= p_n-\frac{(p_{n+1}-p_n)^2}{p_{n+2}-2p_{n+1}+p_n}, \end{aligned}$$

which results in the more rapidly convergent sequence:

$$p \approx p_{n+3} = p_n-\frac{(p_{n+1}-p_n)^2}{p_{n+2}-2p_{n+1}+p_n}.$$

Implementation in MATLAB

Here is the source for an implementation of Steffensen's Method in MATLAB.

```
function Steffensen(f,p0,tol)
% This function takes as inputs: a fixed point iteration function, f,
% and initial guess to the fixed point, p0, and a tolerance, tol.
% The fixed point iteration function is assumed to be input as an
```

```
% inline function.
% This function will calculate and return the fixed point, p,
% that makes the expression f(x) = p true to within the desired
% tolerance, tol.

format compact % This shortens the output.
format long    % This prints more decimal places.

for i=1:1000   % get ready to do a large, but finite, number of itera-
tions.
                % This is so that if the method fails to converge, we
won't
               % be stuck in an infinite loop.
    p1=f(p0);  % calculate the next two guesses for the fixed point.
    p2=f(p1);
    p=p0-(p1-p0)^2/(p2-2*p1+p0) % use Aitken's delta squared method to
                                % find a better approximation to p0.
    if abs(p-p0)<tol  % test to see if we are within tolerance.
        break          % if we are, stop the iterations, we have our
answer.
    end
    p0=p;              % update p0 for the next iteration.
end
if abs(p-p0)>tol       % If we fail to meet the tolerance, we output a
                       % message of failure.
    'failed to converge in 1000 iterations.'
end
```

Implementation in Python

Here is the source for an implementation of Steffensen's Method in Python.

```
def g(f, x: float):
    """First-order divided difference function.
```

```
    Arguments:

        f(callable): Function input to g

        x(float): Point at which to evaluate g

    """

    return lambda x: f(x + f(x)) / f(x) - 1


def steff(f, x: float):

    """Steffenson algorithm for finding roots.


    This recursive generator yields the x_n+1 value first then, when the
generator iterates,

    it yields x_n + 2 from the next level of recursion.


    Arguments:

        f(callable): Function whose root we are searching for

        x(float): Starting value upon first call, each level n that the
function recurses x is x_n

    """


    if g(f, x)(x) != 0:

        yield x - f(x) / g(f, x)(x)  # First give x_n + 1

        yield from steff(f, x - f(x) / g(f, x)(x))  # Then give new it-
erator
```

Generalization

Steffensen's method can also be used to find an input $x = x_\star$ for a different kind of function F that produces output the same as its input: $x_\star = F(x_\star)$ for the special value $x_\star$. Solutions like $x_\star$ are called fixed points. Many such functions can be used to find their own solutions by repeatedly recycling the result back as input, but the rate of convergence can be slow, or the function can fail to converge at all, depending on the individual function. Steffensen's method accelerates this convergence, to make it quadratic.

This method for finding fixed points of a real-valued function has been generalised for functions $F: X \to X$ on a Banach space X. The generalised method assumes that a family of bounded linear operators $\{L(u,v): u,v \in X\}$ associated with u and v can be found to satisfy the condition:

$$F(u) - F(v) = L(u,v)\,(u - v).$$

In the simple form given in the section before, the function f simply takes in and produces real numbers. There, the function g is a divided difference. In the generalized form here, the operator L is the analogue of a divided difference for use in the Banach space. The operator L is equivalent to a matrix whose entries are all functions of vector arguments u and v.

Steffensen's method is then very similar to the Newton's method, except that it uses the divided difference $L(F(x),x)$ instead of the derivative $F'(x)$. It is thus defined by:

$$x_{n+1} = x_n + [I - L(F(x_n), x_n)]^{-1}(F(x_n) - x_n),$$

for $n = 1, 2, 3, \ldots$, and where I is the identity operator.

If the operator L satisfies

$$\left\| L(u,v) - L(x,y) \right\| \le k\left(\left\| u - x \right\| + \left\| v - y \right\| \right)$$

for some constant k, then the method converges quadratically to a fixed point of F if the initial approximation x_0 is 'sufficiently close' to the desired solution $x_\star$, that satisfies $x_\star = F(x_\star)$.

Fast Inverse Square Root

Lighting and reflection calculations (shown here in the first-person shooter *OpenArena*) use the fast inverse square root code to compute angles of incidence and reflection.

Fast inverse square root, sometimes referred to as Fast InvSqrt() or by the hexadecimal constant 0x5F3759DF, is an algorithm that estimates $1/\sqrt{x}$, the reciprocal (or multiplicative inverse) of the square root of a 32-bit floating-point number x in IEEE 754 floating-point format. This operation is used in digital signal processing to normalize a vector, i.e., scale it to length 1. For example, computer graphics programs use inverse square roots to compute angles of incidence and reflection for lighting and shading. The algorithm is best known for its implementation in 1999 in the source code of *Quake III Arena*, a first-person shooter video game that made heavy use of 3D graphics. The algorithm only started appearing on public forums such as Usenet in 2002 or 2003. At the time, it was generally computationally expensive to compute the reciprocal of a floating-point number, especially on a large scale; the fast inverse square root bypassed this step.

The algorithm accepts a 32-bit floating-point number as the input and stores a halved value for later use. Then, treating the bits representing the floating-point number as a 32-bit integer, a logical shift right by one bit is performed and the result subtracted from the number 0x5F3759DF, which is a floating point representation of an approximation of $\sqrt{2^{127}}$. This results in the first approximation of the inverse square root of the input. Treating the bits again as a floating-point number, it runs one iteration of Newton's method, yielding a more precise approximation.

The algorithm was originally attributed to John Carmack, but an investigation showed that the code had deeper roots in both the hardware and software side of computer graphics. Adjustments and alterations passed through both Silicon Graphics and 3dfx Interactive, with Gary Tarolli's implementation for the SGI Indigo as the earliest known use. It is not known how the constant was originally derived, though investigation has shed some light on possible methods.

With subsequent hardware advancements, especially the x86 SSE instruction rsqrtss, this method is not generally applicable to modern computing, though it remains an interesting example both historically and for more limited machines.

Motivation

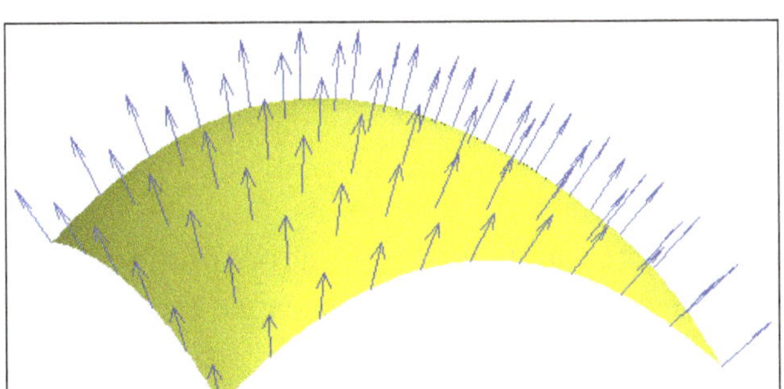

Surface normals are used extensively in lighting and shading calculations, requiring the calculation of norms for vectors. A field of vectors normal to a surface is shown here.

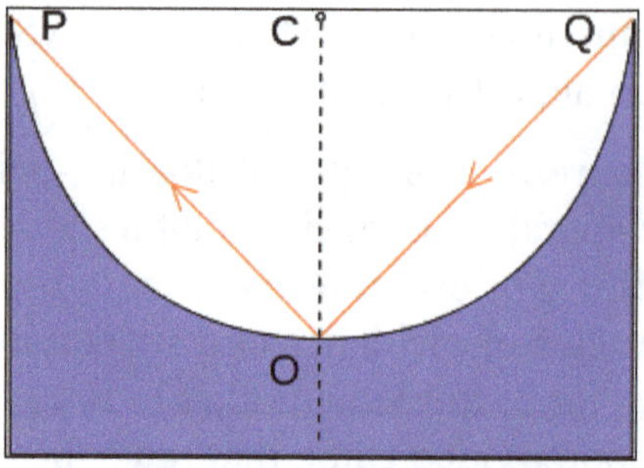

A two-dimensional example of using the normal *C* to find the angle of reflection from the angle of incidence; in this case, on light reflecting from a curved mirror. The fast inverse square root is used to generalize this calculation to three-dimensional space.

The inverse square root of a floating point number is used in calculating a normalized vector. Programs can use normalized vectors to determine angles of incidence and reflection. 3D graphics programs must perform millions of these calculations every second to simulate lighting. When the code was developed in the early 1990s, most floating-point processing power lagged behind the speed of integer processing. This was troublesome for 3D graphics programs before the advent of specialized hardware to handle transform and lighting.

The length of the vector is determined by calculating its Euclidean norm: the square root of the sum of squares of the vector components. When each component of the vector is divided by that length, the new vector will be a unit vector pointing in the same direction. In a 3D graphics program, all vectors are in three-dimensional space, so $\boldsymbol{v}$ would be a vector (v_1, v_2, v_3).

$$\|\mathrm{v}\| = \sqrt{v_1^2 + v_2^2 + v_3^2}$$

is the Euclidean norm of the vector.

$$\hat{\mathrm{v}} = \frac{\mathrm{v}}{\|\mathrm{v}\|}$$

is the normalized (unit) vector, using $||v||^2$ to represent $v_1^2 + v_2^2 + v_3^2$.

$$\hat{\mathrm{v}} = \frac{\mathrm{v}}{\sqrt{\|\mathrm{v}\|^2}}$$

which relates the unit vector to the inverse square root of the distance components. The inverse square root can be used to compute $\hat{\mathrm{v}}$ because this equation is equivalent to

$$\hat{\mathrm{v}} = \mathrm{v}\frac{1}{\sqrt{\|\mathrm{v}\|^2}}$$

where the fraction term is the inverse square root of $||v||^2$.

At the time, floating-point division was generally expensive compared to multiplication; the fast inverse square root algorithm bypassed the division step, giving it its performance advantage. Quake III Arena, a first-person shooter video game, used the fast inverse square root algorithm to accelerate graphics computation, but the algorithm has since been implemented in some dedicated hardware vertex shaders using field-programmable gate arrays (FPGA).

Overview of the Code

The following code is the fast inverse square root implementation from *Quake III Arena*, stripped of C preprocessor directives, but including the exact original comment text:

```
float Q_rsqrt( float number )

{

        long i;

        float x2, y;

        const float threehalfs = 1.5F;


        x2 = number * 0.5F;

        y  = number;

        i  = * ( long * ) &y;                       // evil floating point
bit level hacking

        i  = 0x5f3759df - ( i >> 1 );               // what the fuck?

        y  = * ( float * ) &i;

        y  = y * ( threehalfs - ( x2 * y * y ) );   // 1st iteration

//      y  = y * ( threehalfs - ( x2 * y * y ) );   // 2nd iteration, this
can be removed


        return y;

}
```

At the time, the general method to compute the inverse square root was to calculate an approximation for $1/\sqrt{x}$, then revise that approximation via another method until it came within an acceptable error range of the actual result. Common software methods in the early 1990s drew approximations from a lookup table. The key of the fast inverse square root was to directly compute an approximation by utilizing the structure of floating-point numbers, proving faster than table lookups. The algorithm was approximately four times faster than computing the square root with another method and calculating the reciprocal via floating-point division. The algorithm was designed with the IEEE 754-1985 32-bit floating-point specification in mind, but investigation from Chris Lomont showed that it could be implemented in other floating-point specifications.

The advantages in speed offered by the fast inverse square root kludge came from treating the longword containing the floating point number as an integer then subtracting it from a specific constant, 0x5F3759DF. The purpose of the constant is not immediately clear to someone viewing the code, so, like other such constants found in code, it is often called a magic number. This integer subtraction and bit shift results in a longword which when treated as a floating point number is a rough approximation for the inverse square root of the input number. One iteration of Newton's method is performed to gain some accuracy, and the code is finished. The algorithm generates reasonably accurate results using a unique first approximation for Newton's method; however, it is much slower and less accurate than using the SSE instruction rsqrtss on x86 processors also released in 1999.

In terms of C standards, reinterpreting a floating point value as an integer by dereferencing a casted pointer to it is considered undefined behavior. The proper way to accomplish the fast inverse square root, in a more standard conforming way, is to type-pun floating point values and integers through a union type. Type-punning with a union type is considered undefined behavior in C++ standards.

```
float Q_rsqrt( float number )

{

        const float x2 = number * 0.5F;

        const float threehalfs = 1.5F;


        union {

               float f;

               uint32_t i;

        } conv = {number}; // member 'f' set to value of 'number'.

        conv.i  = 0x5f3759df - ( conv.i >> 1 );

        conv.f  *= ( threehalfs - ( x2 * conv.f * conv.f ) );
```

```
    return conv.f;
}
```

A Worked Example

As an example, the number $x = 0.15625$ can be used to calculate $\frac{1}{\sqrt{x}} \approx 2.52982$. TThe first steps of the algorithm are illustrated here:

```
0011_1110_0010_0000_0000_0000_0000_0000  Bit pattern of both x and i

0001_1111_0001_0000_0000_0000_0000_0000  Shift right one position: (i
>> 1)

0101_1111_0011_0111_0101_1001_1101_1111  The magic number 0x5F3759DF

0100_0000_0010_0111_0101_1001_1101_1111  The result of 0x5F3759DF - (i
>> 1)
```

Interpreting as IEEE 32-bit representation:

`0_01111100_01000000000000000000000` 1.25×2^{-3}

`0_00111110_00100000000000000000000` 1.125×2^{-65}

`0_10111110_01101110101100111011111` $1.432430... \times 2^{63}$

`0_10000000_01001110101100111011111` $1.307430... \times 2^{1}$

Reinterpreting this last bit pattern as a floating point number gives the approximation $y = 2.61486$, which has an error of about 3.4%. After one single iteration of Newton's method, the final result is $y = 2.52549$, an error of only 0.17%.

Algorithm

The algorithm computes $\frac{1}{\sqrt{x}}$ by performing the following steps:

- Alias the argument x to an integer as a way to compute an approximation of $\log_2(x)$.
- Use this approximation to compute an approximation of $\log_2$:

$$\left(\frac{1}{\sqrt{x}}\right) = -1/2\log_2(x)$$

- Alias back to a float, as a way to compute an approximation of the base-2 exponential.
- Refine the approximation using a single iteration of the Newton's method.

Floating-point Representation

Since this algorithm relies heavily on the bit-level representation of single-precision floating-point numbers, a short overview of this representation is provided here. In order to encode a non-zero real number x as a single precision float, the first step is to write x as a normalized binary number:

$$\begin{aligned} x &= \pm 1.b_1b_2b_3\ldots \times 2^{e_x} \\ &= \pm 2^{e_x}(1+m_x) \end{aligned}$$

where the exponent e_x is an integer, $m_x \in [0, 1)$, and $1.b_1b_2b_3\ldots$ is the binary representation of the "significand" $(1 + m_x)$. Since the single bit before the point in the significand is always 1, it need not be stored. From this form, three unsigned integers are computed:

- S_x, the "sign bit", is 0 if $x > 0$, and 1 if $x < 0$ (1 bit).
- $E_x = e_x + B$ is the "biased exponent", where $B = 127$ is the "exponent bias" (8 bits).
- $M_x = m_x \times L$, where $L = 2^{23}$ (23 bits).

These fields are then packed, left to right, into a 32-bit container.

As an example, consider again the number $x = 0.15625 = 0.00101_2$. Normalizing x yields:

$$x = +2^{-3}(1 + 0.25)$$

and thus, the three unsigned integer fields are:

- $S = 0$.
- $E = -3 + 127 = 124 = 01111100_2$.
- $M = 0.25 \times 2^{23} = 2097152 = 01000000000000000000000_2$.

these fields are packed as shown in the figure below:

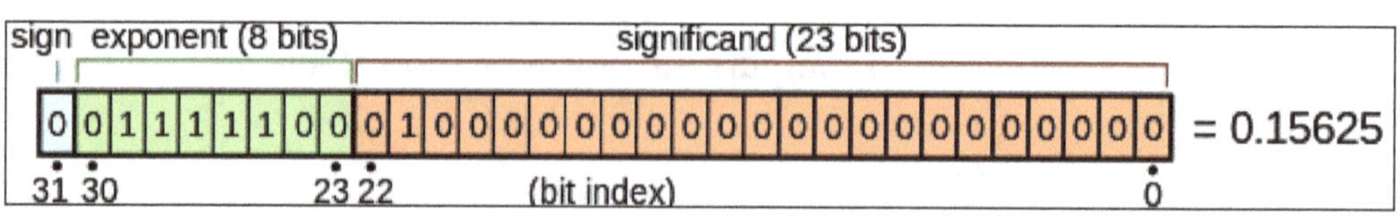

Aliasing to an Integer as an Approximate Logarithm

If $1/\sqrt{x}$ was to be calculated without a computer or a calculator, a table of logarithms would be useful, together with the identity $\log_b\left(1/\sqrt{x}\right) = -\frac{1}{2}\log_b(x)$, which is valid for every base b. The fast inverse square root is based on this identity, and on the fact that aliasing a float32 to an integer gives a rough approximation of its logarithm. Here is how:

If x is a positive normal number:

$$x = 2^{e_x}(1+m_x)$$

then,

$$\log_2(x) = e_x + \log_2(1+m_x)$$

and since $m_x \in [0, 1)$, the logarithm on the right hand side can be approximated by:

$$\log_2(1+m_x) \approx m_x + \sigma$$

where σ is a free parameter used to tune the approximation. For example, $\sigma = 0$ yields exact results at both ends of the interval, while $\sigma \approx 0.0430357$ yields the optimal approximation (the best in the sense of the uniform norm of the error).

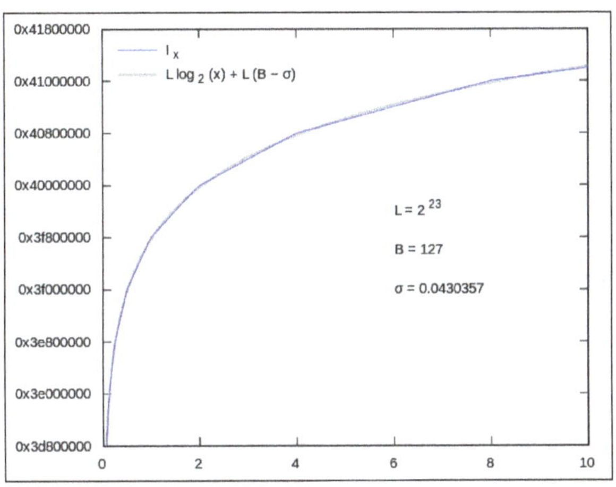

The integer aliased to a floating point number (in blue), compared to a scaled and shifted logarithm (in gray).

Thus there is the approximation:

$$\log_2(x) \approx e_x + m_x + \sigma.$$

Interpreting the floating-point bit-pattern of x as an integer I_x yields:

$$\begin{aligned} I_x &= E_x L + M_x \\ &= L(e_x + B + m_x) \\ &= L(e_x + m_x + \sigma + B - \sigma) \\ &\approx L\log_2(x) + L(B - \sigma). \end{aligned}$$

It then appears that I_x is a scaled and shifted piecewise-linear approximation of $\log_2(x)$, as illustrated in the figure on the right. In other words, $\log_2(x)$ is approximated by:

$$\log_2(x) \approx \frac{I_x}{L} - (B - \sigma).$$

First Approximation of the Result

The calculation of $y = 1/\sqrt{x}$ is based on the identity:

$$\log_2(y) = -\tfrac{1}{2}\log_2(x)$$

Using the approximation of the logarithm above, applied to both x and y, the above equation gives:

$$\frac{I_y}{L} - (B - \sigma) \approx -\frac{1}{2}\left(\frac{I_x}{L} - (B - \sigma)\right)$$

Thus, an approximation of I_y is:

$$I_y \approx \tfrac{3}{2}L(B - \sigma) - \tfrac{1}{2}I_x$$

which is written in the code as:

```
i = 0x5f3759df - ( i >> 1 );
```

The first term above is the magic number:

$$\tfrac{3}{2}L(B - \sigma) = 0\text{x}5\text{F}3759\text{DF}$$

from which it can be inferred that $\sigma \approx 0.0450466$. The second term, $\frac{1}{2}I_x$, is calculated by shifting the bits of I_x one position to the right.

Newton's Method

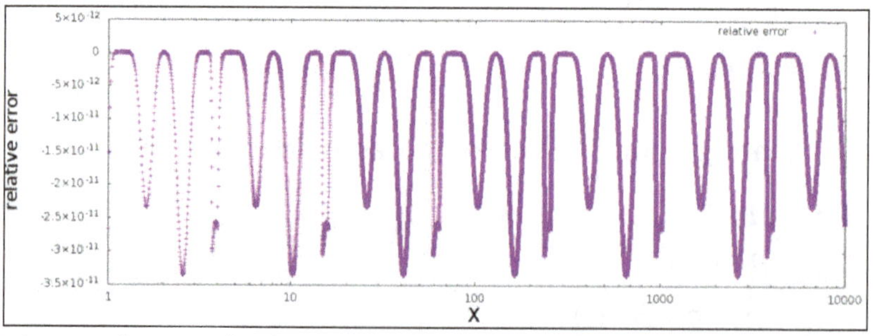

Relative error between fast inverse square root carrying out 3 iterations of Newton's root-finding method, and direct calculation. Note that double precision is adopted.

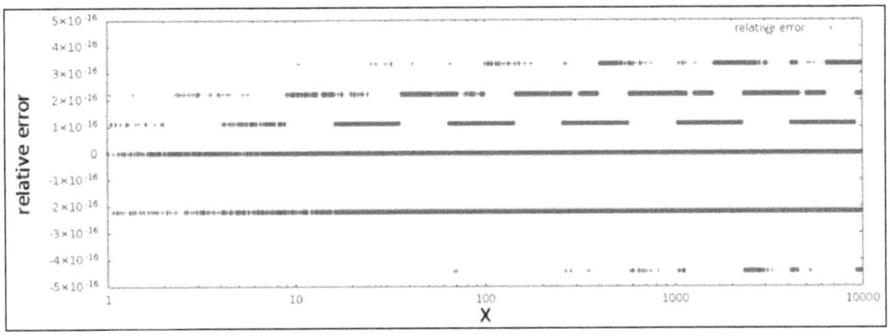

Smallest representable difference between two double precision numbers is reached after carrying out 4 iterations.

With y as the inverse square root, $f(y)=\frac{1}{y^2}-x=0$. The approximation yielded by the earlier steps can be refined by using a root-finding method, a method that finds the zero of a function. The algorithm uses Newton's method: if there is an approximation, y_n for y, then a better approximation y_{n+1} can be calculated by taking $y_n-\frac{f(y_n)}{f'(y_n)}$, where $f'(y_n)$ is the derivative of $f(y)$ at y_n. For the above $f(y)$,

$$y_{n+1}=\frac{y_n\left(3-xy_n^2\right)}{2}$$

where,

$$f(y)=\frac{1}{y^2}-x$$

and,

$$f'(y_n)=-\frac{2}{y^3}.$$

Treating y as a floating-point number, y = y*(threehalfs - x2*y*y); is equivalent to:

$$y_{n+1}=y_n\left(\frac{3}{2}-\frac{x}{2}y_n^2\right)=\frac{y_n\left(3-xy_n^2\right)}{2}.$$

By repeating this step, using the output of the function (y_{n+1}) as the input of the next iteration, the algorithm causes y to converge to the inverse square root. For the purposes of the Quake III engine, only one iteration was used. A second iteration remained in the code but was commented out.

Accuracy

As noted above, the approximation is surprisingly accurate. The graph on the right plots the error of the function (that is, the error of the approximation after it has been improved by running one iteration of Newton's method), for inputs starting at 0.01, where the standard library gives 10.0 as a result, while InvSqrt() gives 9.982522, making the difference 0.017479, or 0.175% of the true value, 10. The absolute error only drops from then on, while the relative error stays within the same bounds across all orders of magnitude.

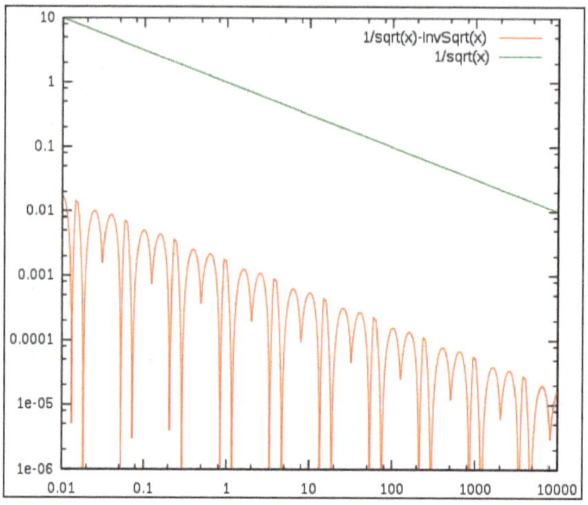

A graph showing the difference between the heuristic fast inverse square root and the direct inversion of square root supplied by libstdc. (Note the log scale on both axes.)

Durand–Kerner Method

In numerical analysis, the Durand–Kerner method, discovered by Karl Weierstrass in 1891 and rediscovered independently by Durand in 1960 and Kerner in 1966, is a root-finding algorithm for solving polynomial equations. In other words, the method can be used to solve numerically the equation:

$$f(x) = 0$$

where f is a given polynomial, which can be taken to be scaled so that the leading coefficient is 1.

Explanation

This explanation considers equations of degree four. It is easily generalized to other degrees.

Let the polynomial f be defined by:

$$f(x) = x^4 + ax^3 + bx^2 + cx + d.$$

for all x.

The known numbers a, b, c, d are the coefficients.

Let the (complex) numbers P,Q,R,S be the roots of this polynomial f.

Then,

$$f(x) = (x-P)(x-Q)(x-R)(x-S)$$

for all x. One can isolate the value P from this equation,

$$P = x - \frac{f(x)}{(x-Q)(x-R)(x-S)}.$$

So if used as a fixed point iteration:

$$x_1 := x_0 - \frac{f(x_0)}{(x_0-Q)(x_0-R)(x_0-S)},$$

it is strongly stable in that every initial point $x_o \neq Q,R,S$ delivers after one iteration the root $P=x_1$.

Furthermore, if one replaces the zeros Q, R and S by approximations $q \approx Q$, $r \approx R$, $s \approx S$, such that q,r,s are not equal to P, then P is still a fixed point of the perturbed fixed point iteration:

$$x_{k+1} := x_k - \frac{f(x_k)}{(x_k-q)(x_k-r)(x_k-s)},$$

since,

$$P - \frac{f(P)}{(P-q)(P-r)(P-s)} = P - 0 = P.$$

Note that the denominator is still different from zero. This fixed point iteration is a contraction mapping for x around P.

The clue to the method now is to combine the fixed point iteration for P with similar iterations for Q,R,S into a simultaneous iteration for all roots.

Initialize p, q, r, s:

$p_0 := (0.4 + 0.9\,i)^0$;

$q_0 := (0.4 + 0.9\,i)^1$;

$r_0 := (0.4 + 0.9\,i)^2$;

$s_0 := (0.4 + 0.9\,i)^3$;

There is nothing special about choosing 0.4 + 0.9 i except that it is neither a real number nor a root of unity.

Make the substitutions for $n = 1,2,3,\cdots$,

$$p_n = p_{n-1} - \frac{f(p_{n-1})}{(p_{n-1}-q_{n-1})(p_{n-1}-r_{n-1})(p_{n-1}-s_{n-1})};$$

$$q_n = q_{n-1} - \frac{f(q_{n-1})}{(q_{n-1}-p_n)(q_{n-1}-r_{n-1})(q_{n-1}-s_{n-1})};$$

$$r_n = r_{n-1} - \frac{f(r_{n-1})}{(r_{n-1}-p_n)(r_{n-1}-q_n)(r_{n-1}-s_{n-1})};$$

$$s_n = s_{n-1} - \frac{f(s_{n-1})}{(s_{n-1}-p_n)(s_{n-1}-q_n)(s_{n-1}-r_n)}.$$

Re-iterate until the numbers *p, q, r, s* essentially stop changing relative to the desired precision. They then have the values *P, Q, R, S* in some order and in the chosen precision. So the problem is solved.

Note that you must use complex number arithmetic, and that the roots are found simultaneously rather than one at a time.

Variations

This iteration procedure, like the Gauss–Seidel method for linear equations, computes one number at a time based on the already computed numbers. A variant of this procedure, like the Jacobi method, computes a vector of root approximations at a time. Both variant are effective root-finding algorithms.

One could also choose the initial values for *p,q,r,s* by some other procedure, even randomly, but in a way that:

- They are inside some not-too-large circle containing also the roots of $f(x)$, e.g. the circle around the origin with radius $1+\max(|a|,|b|,|c|,|d|)$, (where 1,*a,b,c,d* are the coefficients of $f(x)$).

and that:

- They are not too close to each other.

which may increasingly become a concern as the degree of the polynomial increases.

If the coefficients are real and the polynomial has odd degree, then it must have at least one real root. To find this, use a real value of p_0 as the initial guess and make q_0 and r_0, etc, complex conjugate pairs. Then the iteration will preserve these properties; that is, p_n will always be real, and q_n and r_n, etc, will always be conjugate. In this way, the p_n

will converge to a real root P. Alternatively, make all of the initial guesses real; they will remain so.

Example:

This example is from the reference 1992. The equation solved is $x^3 - 3x^2 + 3x - 5 = 0$. The first 4 iterations move p, q, r seemingly chaotically, but then the roots are located to 1 decimal. After iteration number 5 we have 4 correct decimals, and the subsequent iteration number 6 confirms that the computed roots are fixed. This general behaviour is characteristic for the method.

it.-no.	p	q	r
0	+1.0000 + 0.0000i	+0.4000 + 0.9000i	−0.6500 + 0.7200i
1	+1.3608 + 2.0222i	−0.3658 + 2.4838i	−2.3858 − 0.0284i
2	+2.6597 + 2.7137i	+0.5977 + 0.8225i	−0.6320−1.6716i
3	+2.2704 + 0.3880i	+0.1312 + 1.3128i	+0.2821 − 1.5015i
4	+2.5428 − 0.0153i	+0.2044 + 1.3716i	+0.2056 − 1.3721i
5	+2.5874 + 0.0000i	+0.2063 + 1.3747i	+0.2063 − 1.3747i
6	+2.5874 + 0.0000i	+0.2063 + 1.3747i	+0.2063 − 1.3747i

Note that the equation has one real root and one pair of complex conjugate roots, and that the sum of the roots is 3.

Derivation of the Method via Newton's Method

For every n-tuple of complex numbers, there is exactly one monic polynomial of degree n that has them as its zeros (keeping multiplicities). This polynomial is given by multiplying all the corresponding linear factors, that is:

$$g_{\vec{z}}(X) = (X - z_1)\cdots(X - z_n).$$

This polynomial has coefficients that depend on the prescribed zeros,

$$g_{\vec{z}}(X) = X^n + g_{n-1}(\vec{z})X^{n-1} + \cdots + g_0(\vec{z}).$$

Those coefficients are, up to a sign, the elementary symmetric polynomials $\acute{\alpha}_1(\vec{z}),\ldots,\acute{\alpha}_n(\vec{z})$ of degrees $1,\ldots,n$.

To find all the roots of a given polynomial $f(X) = X^n + c_{n-1}X^{n-1} + \cdots + c_0$ with coefficient vector $(c_{n-1},\ldots,c_0)$ simultaneously is now the same as to find a solution vector to the system:

$$\begin{aligned}
c_0 &= g_0(\vec{z}) = (-1)^n \alpha_n(\vec{z}) = (-1)^n z_1 \cdots z_n \\
c_1 &= g_1(\vec{z}) = (-1)^{n-1} \alpha_{n-1}(\vec{z}) \\
&\vdots \\
c_{n-1} &= g_{n-1}(\vec{z}) = -\alpha_1(\vec{z}) = -(z_1 + z_2 + \cdots + z_n).
\end{aligned}$$

The Durand–Kerner method is obtained as the multidimensional Newton's method applied to this system. It is algebraically more comfortable to treat those identities of coefficients as the identity of the corresponding polynomials, $g_{\vec{z}}(X) = f(X)$. In the Newton's method one looks, given some initial vector $\vec{z}$, for an increment vector $\vec{w}$ such that $g_{\vec{z}+\vec{w}}(X) = f(X)$ is satisfied up to second and higher order terms in the increment. For this one solves the identity

$$f(X) - g_{\vec{z}}(X) = \sum_{k=1}^{n} \frac{\partial g_{\vec{z}}(X)}{\partial z_k} w_k = -\sum_{k=1}^{n} w_k \prod_{j \neq k} (X - z_j).$$

If the numbers $z_1, \ldots, z_n$ are pairwise different, then the polynomials in the terms of the right hand side form a basis of the n-dimensional space $\mathbb{C}[X]_{n-1}$ of polynomials with maximal degree $n - 1$. Thus a solution $\vec{w}$ to the increment equation exists in this case. The coordinates of the increment are simply obtained by evaluating the increment equation:

$$-\sum_{k=1}^{n} w_k \prod_{j \neq k} (X - z_j) = f(X) - \prod_{j=1}^{n} (X - z_j)$$

at the points $X = z_k$, which results in:

$$-w_k \prod_{j \neq k} (z_k - z_j) = -w_k g'_{\vec{z}}(z_k) = f(z_k),$$

that is:

$$w_k = -\frac{f(z_k)}{\prod_{j \neq k} (z_k - z_j)}.$$

Root Inclusion via Gerschgorin's Circles

In the quotient ring (algebra) of residue classes modulo $f(X)$, the multiplication by X defines an endomorphism that has the zeros of $f(X)$ as eigenvalues with the corresponding multiplicities. Choosing a basis, the multiplication operator is represented by its coefficient matrix A, the companion matrix of $f(X)$ for this basis.

Since every polynomial can be reduced modulo $f(X)$ to a polynomial of degree $n - 1$ or lower, the space of residue classes can be identified with the space of polynomials of degree bounded by $n - 1$. A problem specific basis can be taken from Lagrange interpolation as the set of n polynomials:

$$b_k(X) = \prod_{1 \leq j \leq n,\, j \neq k} (X - z_j), \quad k = 1, \ldots, n,$$

where $z_1,\ldots,z_n \in \mathbb{C}$ are pairwise different complex numbers. Note that the kernel functions for the Lagrange interpolation are $L_k(X) = \frac{b_k(X)}{b_k(z_k)}$.

For the multiplication operator applied to the basis polynomials one obtains from the Lagrange interpolation:

$$\begin{aligned} X \cdot b_k(X) \bmod f(X) = X \cdot b_k(X) - f(X) &= \sum_{j=1}^{n} \left(z_j \cdot b_k(z_j) - f(z_j) \right) \cdot \frac{b_j(X)}{b_j(z_j)} \\ &= z_k \cdot b_k(X) + \sum_{j=1}^{n} w_j \cdot b_j(X), \end{aligned}$$

where $w_j = -\frac{f(z_j)}{b_j(z_j)}$ are again the Weierstrass updates.

The companion matrix of $f(X)$ is therefore:

$$A = diag(z_1,\ldots,z_n) + \begin{pmatrix} 1 \\ \vdots \\ 1 \end{pmatrix} \cdot (w_1,\ldots,w_n).$$

From the transposed matrix case of the Gershgorin circle theorem it follows that all eigenvalues of A, that is, all roots of $f(X)$, are contained in the union of the disks $D(a_{k,k}, r_k)$ with a radius $r_k = \sum_{j \neq k} |a_{j,k}|$.

Here one has $a_{k,k} = z_k + w_k$, so the centers are the next iterates of the Weierstrass iteration, and radii $r_k = (n-1)|w_k|$ that are multiples of the Weierstrass updates. If the roots of $f(X)$ are all well isolated (relative to the computational precision) and the points $z_1,\ldots,z_n \in \mathbb{C}$ are sufficidently close approximations to these roots, then all the disks will become disjoint, so each one contains exactly one zero. The midpoints of the circles will be better approximations of the zeros.

Every conjugate matrix TAT^{-1} of A is as well a companion matrix of $f(X)$. Choosing T as diagonal matrix leaves the structure of A invariant. The root close to z_k is contained in any isolated circle with center z_k regardless of T. Choosing the optimal diagonal matrix T for every index results in better estimates.

Convergence Results

The connection between the Taylor series expansion and Newton's method suggests that the distance from $z_k + w_k$ to the corresponding root is of the order $O(|w_k|^2)$, if the root is well isolated from nearby roots and the approximation is sufficiently close to the root. So after the approximation is close, Newton's method converges *quadratically*; that is: the error is squared with every step (which will greatly reduce the error once it is less than 1). In the case of the Durand–Kerner method, convergence is quadratic if the vector $\vec{z}=(z_1,\ldots,z_n)$ is close to some permutation of the vector of the roots of f.

For the conclusion of linear convergence there is a more specific result. If the initial vector $\vec{z}$ and its vector of Weierstrass updates $\vec{w}=(w_1,\ldots,w_n)$ satisfies the inequality:

$$\max_{1\le k\le n}|w_k|\le\frac{1}{5n}\min_{1\le j<k\le n}|z_k-z_j|,$$

then this inequality also holds for all iterates, all inclusion disks $D(z_k+w_k,(n-1)|w_k|)$ are disjoint and linear convergence with a contraction factor of *1/2* holds. Further, the inclusion disks can in this case be chosen as:

$$D\left(z_k+w_k,\frac{1}{4}|w_k|\right)\qquad k=1,\ldots,n,$$

each containing exactly one zero of f.

Brent's method is a root-finding algorithm which combines root bracketing, bisection, and inverse quadratic interpolation. It is sometimes known as the van Wijngaarden-Deker-Brent method.

Brent's method uses a Lagrange interpolating polynomial of degree 2. claims that this method will always converge as long as the values of the function are computable within a given region containing a root. Given three points x_1, x_2, and x_3, Brent's method fits x as a quadratic function of y, then uses the interpolation formula:

$$x=\frac{[y-f(x_1)][y-f(x_2)]x_3}{[f(x_3)-f(x_1)][f(x_3)-f(x_2)]}+\frac{[y-f(x_2)][y-f(x_3)]x_1}{[f(x_1)-f(x_2)][f(x_1)-f(x_3)]}+\frac{[y-f(x_3)][y-f(x_1)]x_2}{[f(x_2)-f(x_3)][f(x_2)-f(x_1)]}$$

Subsequent root estimates are obtained by setting $y=0$, giving:

$$x=x_2+\frac{P}{Q},$$

Where:

$$P=S[T(R-T)(x_3-x_2)-(1-R)(x_2-x_1)]$$
$$Q=(T-1)(R-1)(S-1)$$

With,

$$R \equiv \frac{f(x_2)}{f(x_3)}$$

$$S \equiv \frac{f(x_2)}{f(x_1)}$$

$$T \equiv \frac{f(x_1)}{f(x_3)}$$

Numerical Linear Algebra

CHAPTER 4

Numerical linear algebra is the study of matrix operations for creating computer algorithms that provide efficient solutions to mathematical problems. Bartels-Stewart algorithm, GotoBLAS, DADiSP, LAPACK, MATLAB, Armadillo, etc. are studied under its domain. This chapter has been carefully written to provide an easy understanding of numerical linear algebra.

BARTELS–STEWART ALGORITHM

In numerical linear algebra, the Bartels–Stewart algorithm is used to numerically solve the Sylvester matrix equation $AX - XB = C$. Developed by R.H. Bartels and G.W. Stewart in 1971, it was the first numerically stable method that could by systematically applied to solve such equations. The algorithm works by using the real Schur decompositions of *A* and *B* to transform $AX - XB = C$ into a triangular system that can then be solved using forward or backward substitution. In 1979, G. Golub, C. Van Loan and S. Nash introduced an improved version of the algorithm, known as the Hessenberg–Schur algorithm. It remains a standard approach for solving Sylvester equations when *X* is of small to moderate size.

Algorithm

Let $X, C \in \mathbb{R}^{m \times n}$, and assume that the eigenvalues of *A* are distinct from the eigenvalues of *B*. Then, the matrix equation $AX - XB = C$ has a unique solution. The Bartels–Stewart algorithm computes *X* by applying the following steps:

1. Compute the real Schur decompositions.

$$R = U^T A U,$$
$$S = V^T B^T V.$$

The matrices *R* and *S* are block-upper triangular matrices, with square blocks of size no greater than 2.

2. Set $F = U^T C V$.

3. Solve the simplified system $RY - YS^T = F$, where $Y = U^T X V$. This can be done using forward substitution on the blocks. Specifically, if $s_{k-1,k} = 0$, then,

$$(R - s_{kk}I)y_k = f_k + \sum_{j=k+1}^{n} s_{kj}y_j,$$

where y_k is the kth column of Y. When $s_{k-1,k} \neq 0$, columns $[y_{k-1} | y_k]$ should be concatenated and solved for simultaneously.

4. Set $X = UYV^T$.

Computational Cost

Using the QR algorithm, the real Schur decompositions in step 1 require approximately $10(m^3 + n^3)$ flops, so that the overall computational cost is $10(m^3 + n^3) + 2.5(mn^2 + nm^2)$.

Simplifications and Special Cases

In the special case where $B = -A^T$ and C is symmetric, the solution X will also be symmetric. This symmetry can be exploited so that Y is found more efficiently in step 3 of the algorithm.

Hessenberg–Schur Algorithm

The Hessenberg–Schur algorithm replaces the decomposition $R = U^T AU$ in step 1 with the decomposition $H = Q^T AQ$, where H is an upper-Hessenberg matrix. This leads to a system of the form $HY - YS^T = F$ that can be solved using forward substitution. The advantage of this approach is that $H = Q^T AQ$ can be found using Householder reflections at a cost of $(5/3)m^3$ flops, compared to the $10m^3$ flops required to compute the real Schur decomposition of A.

Software and Implementation

The subroutines required for the Hessenberg-Schur variant of the Bartels–Stewart algorithm are implemented in the SLICOT library. These are used in the MATLAB control system toolbox.

Alternative Approaches

For large systems, the $\mathcal{O}(m^3 + n^3)$ cost of the Bartels–Stewart algorithm can be prohibitive. When A and B are sparse or structured, so that linear solves and matrix vector multiplies involving them are efficient, iterative algorithms can potentially perform better. These include projection-based methods, which use Krylov subspace iterations, methods based on the alternating direction implicit (ADI) iteration, and hybridizations that involve both projection and ADI. Iterative methods can also be used to directly construct low rank approximations to X when solving $AX - XB = C$.

ARMADILLO

Armadillo is a linear algebra software library for the C++ programming language. It aims to provide efficient and streamlined base calculations, while at the same time having a straightforward and easy-to-use interface. Its intended target users are scientists and engineers.

It supports integer, floating point (single and double precision), complex numbers, and a subset of trigonometric and statisticsfunctions. Dense and sparse matrices are supported. Various matrix decompositions are provided through optional integration with Linear Algebra PACKage (LAPACK), Automatically Tuned Linear Algebra Software (ATLAS), and ARPACK. High-performance BLAS/LAPACK replacement libraries such as OpenBLAS and Intel MKL can also be used.

The library employs a delayed-evaluation approach (during compile time) to combine several operations into one and reduce (or eliminate) the need for temporaries. Where applicable, the order of operations is optimised. Delayed evaluation and optimisation are achieved through template metaprogramming.

Armadillo is related to the Boost Basic Linear Algebra Subprograms (uBLAS) library, which also uses template metaprogramming. However, Armadillo builds upon ATLAS and LAPACK libraries, thereby providing machine-dependent optimisations and functions not present in uBLAS.

It is open-source software distributed under the permissive Apache License, making it applicable for the development of both open source and proprietary software. The project is supported by the NICTA research centre in Australia.

Example in C++ 11

Here is a trivial example demonstrating Armadillo functionality:

```
// Compile with:

// $ g++ -std=c++11 main.cpp -o file_name -O2 -larmadillo


#include <iostream>

#include <armadillo>

#include <cmath>


int main()

{
```

```
                                                   //     ^
  // Position of a particle                        //     |
  arma::vec Pos = {{0},                            //     | (0,1)
                   {1}};                           //     +---x-->

  // Rotation matrix
  double phi = -3.1416/2;
  arma::mat RotM = {{+cos(phi), -sin(phi)},
                    {+sin(phi), +cos(phi)}};

  Pos.print("Current position of the particle:");
  std::cout << "Rotating the point " << phi*180/3.1416 << " deg" <<
std::endl;

  Pos = RotM*Pos;

  Pos.print("New position of the particle:");      //     ^
                                                   //     x (1,0)
                                                   //     |
                                                   //     +------>

  return 0;
}
```

Example in C++ 98

```
#include <iostream>
#include <armadillo>
```

```
int main()

{

  arma::vec b;

  b << 2.0 << 5.0 << 2.0;


  // arma::endr represents the end of a row in a matrix

  arma::mat A;

  A << 1.0 << 2.0 << arma::endr

    << 2.0 << 3.0 << arma::endr

    << 1.0 << 3.0 << arma::endr;


  std::cout << "Least squares solution:\n";

  std::cout << arma::solve(A,b) << '\n';


  return 0;

}
```

AUTOMATICALLY TUNED LINEAR ALGEBRA SOFTWARE

Automatically Tuned Linear Algebra Software (ATLAS) is a software library for linear algebra. It provides a mature open source implementation of BLAS APIs for C and Fortran77.

ATLAS is often recommended as a way to automatically generate an optimized BLAS library. While its performance often trails that of specialized libraries written for one specific hardware platform, it is often the first or even only optimized BLAS implementation available on new systems and is a large improvement over the generic BLAS available at Netlib. For this reason, ATLAS is sometimes used as a performance baseline for comparison with other products.

ATLAS runs on most Unix-like operating systems and on Microsoft Windows (using Cygwin). It is released under a BSD-style license without advertising clause, and many well-known mathematics applications including MATLAB, Mathematica, Scilab, SageMath, and some builds of GNU Octave may use it.

Functionality

ATLAS provides a full implementation of the BLAS APIs as well as some additional functions from LAPACK, a higher-level library built on top of BLAS. In BLAS, functionality is divided into three groups called levels 1, 2 and 3.

- Level 1 contains *vector operations* of the form:

  $$\mathrm{y} \leftarrow \alpha \mathrm{x} + \mathrm{y}$$

  as well as scalar dot products and vector norms, among other things.

- Level 2 contains *matrix-vector operations* of the form:

  $$\mathrm{y} \leftarrow \alpha A \mathrm{x} + \beta \mathrm{y}$$

  as well as solving $T\mathrm{x} = \mathrm{y}$ for x with T being triangular, among other things.

- Level 3 contains *matrix-matrix operations* such as the widely used General Matrix Multiply (GEMM) operation:

  $$C \leftarrow \alpha AB + \beta C$$

  as well as solving $B \leftarrow \alpha T^{-1} B$ for triangular matrices T, among other things.

Optimization Approach

The optimization approach is called Automated Empirical Optimization of Software (AEOS), which identifies four fundamental approaches to computer assisted optimization of which ATLAS employs three:

- Parameterization: Searching over the parameter space of a function, used for blocking factor, cache edge, etc.

- Multiple Implementation: Searching through various approaches to implementing the same function, e.g., for SSE support before intrinsics made them available in C code.

- Code generation: Programs that write programs incorporating what knowledge they can about what will produce the best performance for the system.
- Optimization of the level 1 BLAS uses parameterization and multiple implementation.

  Every ATLAS level 1 BLAS function has its own kernel. Since it would be difficult to maintain thousands of cases in ATLAS there is little architecture specific optimization for Level 1 BLAS. Instead multiple implementation is relied upon to allow for compiler optimization to produce high performance implementation for the system.
- Optimization of the level 2 BLAS uses parameterization and multiple implementation.

  With N^2 data and N^2 operations to perform the function is usually limited by bandwidth to memory, and thus there is not much opportunity for optimization

  All routines in the ATLAS level 2 BLAS are built from two Level 2 BLAS kernels:
  - GEMV—matrix by vector multiply update:

    $\mathbf{y} \leftarrow \alpha A\mathbf{x} + \beta\mathbf{y}$
  - GER—general rank 1 update from an outer product:

    $A \leftarrow \alpha \mathrm{xy}^T + A$
  - Optimization of the level 3 BLAS uses code generation and the other two techniques.

  Since we have N^3 ops with only N^2 data, there are many opportunities for optimization.

Level 3 BLAS

Most of the Level 3 BLAS is derived from GEMM, so that is the primary focus of the optimization.

$O(n^3)$ Operations vs. $O(n^2)$ Data

The intuition that the n^3 operations will dominate over the n^2 data accesses only works for roughly square matrices. The real measure should be some kind of surface area to volume. The difference becomes important for very non-square matrices.

Copying the Inputs

Copying the inputs allows the data to be arranged in a way that provides optimal access for the kernel functions, but this comes at the cost of allocating temporary space, and an extra read and write of the inputs.

So the first question GEMM faces is, can it afford to copy the inputs?

If so,

- Put into block major format with good alignment.
- Take advantage of user contributed kernels and cleanup.
- Handle the transpose cases with the copy: make everything into TN (transpose - no-transpose).
- Deal with α in the copy.

If not,

- Use the nocopy version.
- Make no assumptions on the stride of matrix A and B in memory.
- Handle all transpose cases explicitly.
- No guarantee about alignment of data.
- Support α specific code.
- Run the risk of TLB issues, bad strides, etc.

The actual decision is made through a simple heuristic which checks for "skinny cases".

Cache Edge

For 2nd Level Cache blocking a single cache edge parameter is used. The high level choose an order to traverse the blocks: *ijk, jik, ikj, jki, kij, kji*. These need not be the same order as the product is done within a block.

Typically chosen orders are *ijk* or *jik*. For *jik* the ideal situation would be to copy A and the NB wide panel of B. For *ijk* swap the role of A and B.

Choosing the bigger of M or N for the outer loop reduces the footprint of the copy. But for large K ATLAS does not even allocate such a large amount of memory. Instead it defines a parameter, Kp, to give best use of the L2 cache. Panels are limited to Kp in length. It first tries to allocate (in the *jik* case) $M \cdot p + NB \cdot Kp + NB \cdot NB$. If that fails it tries $2 \cdot Kp \cdot NB + NB \cdot NB$. (If that fails it uses the no-copy version of GEMM, but this case is unlikely for reasonable choices of cache edge.) Kp is a function of cache edge and NB.

LAPACK

When integrating the ATLAS BLAS with LAPACK an important consideration is the choice of blocking factor for LAPACK. If the ATLAS blocking factor is small enough the blocking factor of LAPACK could be set to match that of ATLAS.

To take advantage of recursive factorization, ATLAS provides replacement routines for some LAPACK routines. These simply overwrite the corresponding LAPACK routines from Netlib.

Need for Installation

Installing ATLAS on a particular platform is a challenging process which is typically done by a system vendor or a local expert and made available to a wider audience.

For many systems, architectural default parameters are available; these are essentially saved searches plus the results of hand tuning. If the arch defaults work they will likely get 10-15% better performance than the install search. On such systems the installation process is greatly simplified.

GOTOBLAS

In scientific computing, GotoBLAS and GotoBLAS2 are open source implementations of the BLAS (Basic Linear Algebra Subprograms) API with many hand-crafted optimizations for specific processor types. GotoBLAS was developed by Kazushige Goto at the Texas Advanced Computing Center. As of 2003, it was used in seven of the world's ten fastest supercomputers.

GotoBLAS remains available, but development ceased with a final version touting optimal performance on Intel's Nehalem architecture. OpenBLAS is an actively maintained fork of GotoBLAS, developed at the Lab of Parallel Software and Computational Science, ISCAS.

GotoBLAS was written by Goto during his sabbatical leave from the Japan Patent Office in 2002. It was initially optimized for the Pentium 4 processor and managed to immediately boost the performance of a supercomputer based on that CPU from 1.5 TFLOPS to 2 TFLOPS. As of 2005, the library was available at no cost for non-commercial use. A later open source version was released under the terms of the BSD license.

GotoBLAS's matrix-matrix multiplication routine, called GEMM in BLAS terms, is highly tuned for the x86 and AMD64 processor architectures by means of handcrafted assembly code. It follows a similar decomposition into smaller "kernel" routines that other BLAS implementations use, but where earlier implementations streamed data from the L1 processor cache, GotoBLAS uses the L2 cache. The kernel used for GEMM is a routine called GEBP, for "General block-times-panel multiply", which was experimentally found to be "inherently superior" over several other kernels that were considered in the design.

Several other BLAS routines are, as is customary in BLAS libraries, implemented in terms of GEMM.

DADISP

DADiSP (Data Analysis and Display, pronounced day-disp) is a numerical computing environment developed by DSP Development Corporation which allows one to display and manipulate data series, matrices and images with an interface similar to a spreadsheet. DADiSP is used in the study of signal processing, numerical analysis, statistical and physiological data processing.

Interface

DADiSP is designed to perform technical data analysis in a spreadsheet like environment. However, unlike a typical business spreadsheet that operates on a table of cells each of which contain single scalar values, a DADiSP Worksheet consists of multiple interrelated windows where each window contains an entire series or multi-column matrix. A window not only stores the data, but also displays the data in several interactive forms, including 2D graphs, XYZ plots, 3D surfaces, images and numeric tables.

Like a traditional spreadsheet, the windows are linked such that a change to the data in one window automatically updates all dependent windows both numerically and graphically. Users manipulate data primarily through windows. A DADiSP window is normally referred to by the letter "W" followed by a window number, as in "W1". For example, the formula $W1$: 1..3 assigns the series values {1, 2, 3} to "W1". The formula $W2$: $W1*W1$ sets a second window to compute the square of each value in "W1" such that "W2" will contain the series {1, 4, 9}. If the values of "W1" change to {3, 5, 2, 4}, the values of "W2" automatically update to {9, 25, 4, 16}.

Programming Language

DADiSP includes a series based programming language called SPL (Series Processing Language) used to implement custom algorithms. SPL has a C/C++ like syntax and is incrementally compiled into intermediate bytecode, which is executed by a virtual machine. SPL supports both standard variables assigned with = and "hot" variables assigned with :=. For example, the statement $A = 1..3$ assigns the series {1, 2, 3} to the standard variable "A". The square of the values can be assigned with $B = A * A$. Variable "B" contains the series {1, 3, 9}. If "A" changes, "B" does *not* change because "B" preserves the values as assigned without regard to the future state of "A".

However, the statement $A := 1..3$ creates a "hot" variable. A hot variable is analogous to a window, except hot variables do not display their data. The assignment $B := A * A$ computes the square of the values of "A" as before, but now if "A" changes, "B" automatically updates. Setting $A = \{3, 5, 2, 4\}$ causes "B" to automatically update with {9, 25, 4, 16}.

EIGENMODE EXPANSION

Eigenmode expansion (EME) is a computational electrodynamics modelling technique. It is also referred to as the mode matching technique or the bidirectional eigenmode propagation method (BEP method). Eigenmode expansion is a linear frequency-domain method.

It offers very strong benefits compared with FDTD, FEM and the beam propagation method for the modelling of optical waveguides, and it is a popular tool for the modelling linear effects in fiber optics and silicon photonics devices.

Principles of the EME Method

Eigenmode expansion is a rigorous technique to simulate electromagnetic propagation which relies on the decomposition of the electromagnetic fields into a basis set of local eigenmodes that exists in the cross section of the device. The eigenmodes are found by solving Maxwell's equations in each local cross-section. The method can be fully vectorial provided that the mode solvers themselves are fully vectorial.

In a typical waveguide, there are a few guided modes (which propagate without coupling along the waveguide) and an infinite number of radiation modes (which carry optical power away from the waveguide). The guided and radiation modes together form a complete basis set. Many problems can be resolved by considering only a modest number of modes, making EME a very powerful method.

As can be seen from the mathematical formulation, the algorithm is inherently bi-directional. It uses the scattering matrix (S-matrix) technique to join different sections of the waveguide or to model nonuniform structures. For structures that vary continuously along the z-direction, a form of z-discretisation is required. Advanced algorithms have been developed for the modelling of optical tapers.

Mathematical Formulation

In a structure where the optical refractive index does not vary in the z direction, the solutions of Maxwell's equations take the form of a plane wave:

$$E(x,y,z) = E(x,y)e^{(i\beta z)}$$

We assume here a single wavelength and time dependence of the form $\exp(i\omega t)$.

Mathematically $E(x,y)e^{(i\beta z)}$ and β are the eigenfunction and eigenvalues of maxwell's equations for conditions with simple harmonic z-dependence.

We can express any solution of Maxwell's equations in terms of a superposition of the forward and backward propagating modes:

$$E(x,y,z) = \sum_{k=1}^{M}(a_k e^{(i\beta_k z)} + b_k e^{(-i\beta_k z)})E_k(x,y)$$

$$H(x,y,z)=\sum_{k=1}^{M}(a_k e^{(i\beta_k z)}-b_k e^{(-i\beta_k z)})H_k(x,y)$$

These equations provide a rigorous solution of Maxwell's equations in a linear medium, the only limitation being the finite number of modes.

When there is a change in the structure along the z-direction, the coupling between the different input and output modes can be obtained in the form of a scattering matrix. The scattering matrix of a discrete step can be obtained rigorously by applying the boundary conditions of Maxwell's equations at the interface; this requires calculation of the modes on both sides of the interface and their overlaps. For continuously varying structures (e.g. tapers), the scattering matrix can be obtained by discretising the structure along the z-axis.

Strengths of the EME Method

- The EME method is ideal for the modelling of guided optical components, for fibre and integrated geometries. The mode calculation can take advantage of symmetries of the structure; for instance cylindrically symmetric structures can be modelled very efficiently.
- The method is fully vectorial (provided that it relies on a fully vectorial mode solver) and fully bidirectional.
- As it relies on a scattering matrix approach, all reflections are taken into account.
- Unlike the beam propagation method, which is only valid under the slowly varying envelope approximation, eigenmode expansion provides a rigorous solution to Maxwell's equations.
- It is generally much more efficient than FDTD or FEM as it does not require fine discretisation (i.e. on the scale of the wavelength) along the direction of propagation.
- The scattering matrix approach provides a flexible calculation framework, potentially allowing users to only re-calculate modified parts of the structure when performing parameter scan studies.
- It is an excellent technique to model long devices or devices composed of metals.
- Fully analytical solutions can be obtained for the modelling of 1D+Z structures.

Limitations of the EME Method

- EME is limited to linear problems; nonlinear problems may be modelled using iterative techniques.

- EME may be inefficient for modeling structures requiring a very large number of modes, which limits the size of the cross-section for 3D problems.

IN-PLACE MATRIX TRANSPOSITION

In-place matrix transposition, also called in-situ matrix transposition, is the problem of transposing an $N \times M$ matrix in-place in computer memory, ideally with $O(1)$ (bounded) additional storage, or at most with additional storage much less than NM. Typically, the matrix is assumed to be stored in row-major order or column-major order (i.e., contiguous rows or columns, respectively, arranged consecutively).

Performing an in-place transpose (in-situ transpose) is most difficult when $N \neq M$, i.e. for a non-square (rectangular) matrix, where it involves a complicated permutation of the data elements, with many cycles of length greater than 2. In contrast, for a square matrix ($N = M$), all of the cycles are of length 1 or 2, and the transpose can be achieved by a simple loop to swap the upper triangle of the matrix with the lower triangle. Further complications arise if one wishes to maximize memory locality in order to improve cache line utilization or to operate out-of-core(where the matrix does not fit into main memory), since transposes inherently involve non-consecutive memory accesses.

The problem of non-square in-place transposition has been studied since at least the late 1950s, and several algorithms are known, including several which attempt to optimize locality for cache, out-of-core, or similar memory-related contexts.

On a computer, one can often avoid explicitly transposing a matrix in memory by simply accessing the same data in a different order. For example, software libraries for linear algebra, such as BLAS, typically provide options to specify that certain matrices are to be interpreted in transposed order to avoid data movement.

However, there remain a number of circumstances in which it is necessary or desirable to physically reorder a matrix in memory to its transposed ordering. For example, with a matrix stored in row-major order, the rows of the matrix are contiguous in memory and the columns are discontiguous. If repeated operations need to be performed on the columns, for example in a fast Fourier transform algorithm, transposing the matrix in memory (to make the columns contiguous) may improve performance by increasing memory locality. Since these situations normally coincide with the case of very large matrices (which exceed the cache size), performing the transposition in-place with minimal additional storage becomes desirable.

Also, as a purely mathematical problem, in-place transposition involves a number of interesting number theory puzzles that have been worked out over the course of several decades.

Example:

For example, consider the 2×4 matrix:

$$\begin{bmatrix} 11 & 12 & 13 & 14 \\ 21 & 22 & 23 & 24 \end{bmatrix}.$$

In row-major format, this would be stored in computer memory as the sequence (11, 12, 13, 14, 21, 22, 23, 24), i.e. the two rows stored consecutively. If we transpose this, we obtain the 4×2 matrix:

$$\begin{bmatrix} 11 & 21 \\ 12 & 22 \\ 13 & 23 \\ 14 & 24 \end{bmatrix}$$

which is stored in computer memory as the sequence (11, 21, 12, 22, 13, 23, 14, 24).

Position	0	1	2	3	4	5	6	7
Original storage	11	12	13	14	21	22	23	24
Transposed storage	11	21	12	22	13	23	14	24

If we number the storage locations 0 to 7, from left to right, then this permutation consists of four cycles:

(0), (1 2 4), (3 6 5), (7)

That is, the value in position 0 goes to position 0 (a cycle of length 1, no data motion). Next, the value in position 1 (in the original storage: 11, 12, 13, 14, 21, 22, 23, 24) goes to position 2 (in the transposed storage 11, 21, 12, 22, 13, 23, 14, 24), while the value in position 2 (11, 12, 13, 14, 21, 22, 23, 24) goes to position 4 (11, 21, 12, 22, 13, 23, 14, 24), and position 4 (11, 12, 13, 14, 21, 22, 23, 24) goes back to position 1 (11, 21, 12, 22, 13, 23, 14, 24). Similarly for the values in position 7 and positions (3 6 5).

Properties of the Permutation

In the following, we assume that the $N\times M$ matrix is stored in row-major order with zero-based indices. This means that the (n,m) element, for $n = 0,\ldots,N-1$ and $m = 0,\ldots,M-1$, is stored at an address $a = Mn + m$ (plus some offset in memory, which we ignore). In the transposed $M\times N$ matrix, the corresponding (m,n) element is stored at the address $a' = Nm + n$, again in row-major order. We define the transposition permutation to be the function $a' = P(a)$ such that:

$$Nm+n = P(Mn+m) \text{ for all } (n,m)\in[0,N-1]\times[0,M-1].$$

This defines a permutation on the numbers $a = 0,\ldots, MN-1$.

It turns out that one can define simple formulas for *P* and its inverse. First:

$$P(a) = \begin{cases} MN-1 & \textit{if } a = MN-1, \\ Na \bmod MN-1 & \textit{otherwise}, \end{cases}$$

where "mod" is the modulo operation.

If $0 \le a = Mn + m < MN - 1$, then $Na \bmod (MN-1) = MN\,n + Nm \bmod (MN-1) = n + Nm$.

Second, the inverse permutation is given by:

$$P^{-1}(a') = \begin{cases} MN-1 & \textit{if } a' = MN-1, \\ Ma' \bmod MN-1 & \textit{otherwise}. \end{cases}$$

This is just a consequence of the fact that the inverse of an $N{\times}M$ transpose is an $M{\times}N$ transpose, although it is also easy to show explicitly that P^{-1} composed with *P* gives the identity.

As proved by Cate and Twigg, the number of fixed points (cycles of length 1) of the permutation is precisely 1 + gcd(N–1, M–1), where gcd is the greatest common divisor. For example, with $N = M$ the number of fixed points is simply *N* (the diagonal of the matrix). If $N - 1$ and $M - 1$ are coprime, on the other hand, the only two fixed points are the upper-left and lower-right corners of the matrix.

The number of cycles of any length $k>1$ is given by:

$$\frac{1}{k}\sum_{d|k}\mu(k/d)\gcd(N^d-1, MN-1),$$

where μ is the Möbius function and the sum is over the divisors *d* of *k*.

Furthermore, the cycle containing a=1 (i.e. the second element of the first row of the matrix) is always a cycle of maximum length *L*, and the lengths *k* of all other cycles must be divisors of *L*.

For a given cycle *C*, every $x \in C$ element has the same greatest common divisor $d = \gcd(x, MN-1)$.

Let s be the smallest element of the cycle, and $d = \gcd(s, MN-1)$. From the definition of the permutation *P* above, every other element *x* of the cycle is obtained by repeatedly multiplying s by *N* modulo MN–1, and therefore every other element is divisible by *d*. But, since *N* and $MN - 1$ are coprime, *x* cannot be divisible by any factor of $MN - 1$ larger than d, and hence $d = \gcd(x, MN-1)$.

This theorem is useful in searching for cycles of the permutation, since an efficient search can look only at multiples of divisors of $MN-1$.

Laflin and Brebner pointed out that the cycles often come in pairs, which is exploited by several algorithms that permute pairs of cycles at a time. In particular, let s be the smallest element of some cycle C of length k. It follows that $MN-1-s$ is also an element of a cycle of length k (possibly the same cycle).

Proof by the Definition of *P* above

The length k of the cycle containing s is the smallest $k > 0$ such that $sN^k = s \bmod (MN-1)$. Clearly, this is the same as the smallest $k > 0$ such that $(-s)N^k = -s \bmod (MN-1)$, since we are just multiplying both sides by -1, and $MN-1-s = -s \bmod (MN-1)$.

Algorithms

The following briefly summarizes the published algorithms to perform in-place matrix transposition.

Accessor Transpose

Because physically transposing a matrix is computationally expensive, instead of moving values in memory, the access path may be transposed instead. It is trivial to perform this operation for CPU access, as the access paths of iterators must simply be exchanged, however hardware acceleration may require that still be physically re-aligned.

Square Matrices

For a square $N \times N$ matrix $A_{n,m} = A(n,m)$, in-place transposition is easy because all of the cycles have length 1 (the diagonals $A_{n,n}$) or length 2 (the upper triangle is swapped with the lower triangle). Pseudocode to accomplish this (assuming zero-based array indices) is:

```
for n = 0 to N - 2

    for m = n + 1 to N - 1

        swap A(n,m) with A(m,n)
```

This type of implementation, while simple, can exhibit poor performance due to poor cache-line utilization, especially when N is a power of two (due to cache-line conflicts in a CPU cache with limited associativity). The reason for this is that, as m is incremented in the inner loop, the memory address corresponding to $A(n,m)$ or $A(m,n)$ jumps discontiguously by N in memory (depending on whether the array is in column-major or row-major format, respectively). That is, the algorithm does not exploit locality of reference.

One solution to improve the cache utilization is to "block" the algorithm to operate on several numbers at once, in blocks given by the cache-line size; unfortunately, this means that the algorithm depends on the size of the cache line (it is "cache-aware"), and on a modern computer with multiple levels of cache it requires multiple levels of machine-dependent blocking. Instead, it has been suggested that better performance can be obtained by a recursive algorithm: divide the matrix into four submatrices of roughly equal size, transposing the two submatrices along the diagonal recursively and transposing and swapping the two submatrices above and below the diagonal. (When N is sufficiently small, the simple algorithm above is used as a base case, as naively recurring all the way down to N=1 would have excessive function-call overhead.) This is a cache-oblivious algorithm, in the sense that it can exploit the cache line without the cache-line size being an explicit parameter.

Non-square Matrices: Following the Cycles

For non-square matrices, the algorithms are more complicated. Many of the algorithms prior to 1980 could be described as "follow-the-cycles" algorithms. That is, they loop over the cycles, moving the data from one location to the next in the cycle. In pseudo-code form:

```
for each length>1 cycle C of the permutation
    pick a starting address s in C
    let D = data at s
    let x = predecessor of s in the cycle
    while x ≠ s
        move data from x to successor of x
        let x = predecessor of x
    move data from D to successor of s
```

The differences between the algorithms lie mainly in how they locate the cycles, how they find the starting addresses in each cycle, and how they ensure that each cycle is moved exactly once. Typically, the cycles are moved in pairs, since s and $MN-1-s$ are in cycles of the same length (possibly the same cycle). Sometimes, a small scratch array, typically of length $M+N$ is used to keep track of a subset of locations in the array that have been visited, to accelerate the algorithm.

In order to determine whether a given cycle has been moved already, the simplest scheme would be to use $O(MN)$ auxiliary storage, one bit per element, to indicate whether a given element has been moved. To use only $O(M+N)$ or even $O(\log MN)$ auxiliary storage, more complicated algorithms are required, and the known algorithms have a worst-case linearithmic computational cost of $O(MN \log MN)$ at best, as first proved by Knuth.

Such algorithms are designed to move each data element exactly once. However, they also involve a considerable amount of arithmetic to compute the cycles, and require heavily non-consecutive memory accesses since the adjacent elements of the cycles differ by multiplicative factors of N.

Improving Memory Locality at the Cost of Greater Total Data Movement

Several algorithms have been designed to achieve greater memory locality at the cost of greater data movement, as well as slightly greater storage requirements. That is, they may move each data element more than once, but they involve more consecutive memory access (greater spatial locality), which can improve performance on modern CPUs that rely on caches, as well as on SIMD architectures optimized for processing consecutive data blocks. The oldest context in which the spatial locality of transposition seems to have been studied is for out-of-core operation, where the matrix is too large to fit into main memory ("core").

For example, if $d = \gcd(N,M)$ is not small, one can perform the transposition using a small amount (NM/d) of additional storage, with at most three passes over the array. Two of the passes involve a sequence of separate, small transpositions (which can be performed efficiently out of place using a small buffer) and one involves an in-place $d \times d$ square transposition of NM / d^2 blocks (which is efficient since the blocks being moved are large and consecutive, and the cycles are of length at most 2). This is further simplified if N is a multiple of M (or vice versa), since only one of the two out-of-place passes is required.

Another algorithm for non-coprime dimensions, involving multiple subsidiary transpositions, was described by Catanzaro et al. For the case where $|N - M|$ is small, Dow describes another algorithm requiring $|N-M| \cdot min(N,M)$ additional storage, involving a $min(N,M) \cdot min(N,M)$ square transpose preceded or followed by a small out-of-place transpose. Frigo & Johnson describe the adaptation of these algorithms to use cache-oblivious techniques for general-purpose CPUs relying on cache lines to exploit spatial locality.

Work on out-of-core matrix transposition, where the matrix does not fit in main memory and must be stored largely on a hard disk, has focused largely on the $N = M$ square-matrix case, with some exceptions.

JULIA

Julia is a high-level, high-performance, dynamic programming language. While it is a general purpose language and can be used to write any application, many of its features are well-suited for high-performance numerical analysis and computational science.

Distinctive aspects of Julia's design include a type system with parametric polymorphism in a dynamic programming language; with multiple dispatch as its core programming paradigm. Julia supports concurrent, (composable) paralleland distributed computing (with or without using MPI and/or the built-in corresponding to "OpenMP-style" threads), and direct calling of C and Fortran libraries without glue code. A just-in-time compiler that is referred to as "just-ahead-of-time" in the Julia community is used.

Julia is garbage-collected, uses eager evaluation, and includes efficient libraries for floating-point calculations, linear algebra, random number generation, and regular expression matching. Many libraries are available, including some (e.g., for fast Fourier transforms) that were previously bundled with Julia and are now separate.

Tools available for Julia include IDEs; with integrated tools, e.g. a linter, profiler (and flame graph support available for the built-in one), debugger, and the Rebugger.jl package "supports repeated-execution debugging" and more.

Language Features

Though designed for numerical computing, Julia is a general-purpose programming language. It is also useful for low-level systems programming, as a specification language, and for web programming at both server and client side.

The main features of Julia are:

- Multiple dispatch: Providing ability to define function behavior across many combinations of argument types.
- Dynamic type system: Types for documentation, optimization, and dispatch.
- Good performance, approaching that of statically-typed languages like C.
- A built-in package manager.
- Lisp-like macros and other metaprogramming facilities.
- Call Python functions: Use the PyCall package.
- Call C functions directly: No wrappers or special APIs.
- Powerful shell-like abilities to manage other processes.
- Designed for parallel and distributed computing.
- Coroutines: Lightweight *green* threading.
- User-defined types are as fast and compact as built-ins.

- Automatic generation of efficient, specialized code for different argument types.
- Elegant and extensible conversions and promotions for numeric and other types.
- Efficient support for Unicode, including but not limited to UTF-8.

Multiple dispatch (also termed multimethods in Lisp) is a generalization of single dispatch – the polymorphic mechanism used in common object-oriented programming (OOP) languages – that uses inheritance. In Julia, all concrete types are subtypes of abstract types, directly or indirectly subtypes of the *Any* type, which is the top of the type hierarchy. Concrete types can not themselves be subtyped the way they can in other languages; composition is used instead.

Julia draws significant inspiration from various dialects of Lisp, including Scheme and Common Lisp, and it shares many features with Dylan, also a multiple-dispatch-oriented dynamic language (which features an ALGOL-like free-form infix syntax rather than a Lisp-like prefix syntax, while in Julia "everything" is an expression), and with Fortress, another numerical programming language (which features multiple dispatch and a sophisticated parametric type system). While Common Lisp Object System (CLOS) adds multiple dispatch to Common Lisp, not all functions are generic functions.

In Julia, Dylan, and Fortress extensibility is the default, and the system's built-in functions are all generic and extensible. In Dylan, multiple dispatch is as fundamental as it is in Julia: all user-defined functions and even basic built-in operations like + are generic. Dylan's type system, however, does not fully support parametric types, which are more typical of the ML lineage of languages. By default, CLOS does not allow for dispatch on Common Lisp's parametric types; such extended dispatch semantics can only be added as an extension through the CLOS Metaobject Protocol. By convergent design, Fortress also features multiple dispatch on parametric types; unlike Julia, however, Fortress is statically rather than dynamically typed, with separate compiling and executing phases. The language features are summarized in the following table:

Language	Type system	Generic functions	Parametric types
Julia	Dynamic	Default	Yes
Common Lisp	Dynamic	Opt-in	Yes (but no dispatch)
Dylan	Dynamic	Default	Partial (no dispatch)
Fortress	Static	Default	Yes

By default, the Julia runtime must be pre-installed as user-provided source code is run. Alternatively, a standalone executable that needs no Julia source code can be built with ApplicationBuilder.jl and PackageCompiler.jl.

Julia's syntactic macros (used for metaprogramming), like Lisp macros, are more powerful than text-substitution macros used in the pre-processor of some other languages such as C, because they work at the level of abstract syntax trees (ASTs). Julia's macro system is hygienic, but also supports deliberate capture when desired (like for anaphoric macros) using the *esc* construct.

Interaction

The Julia official distribution includes a "full-featured interactive command-line REPL" (read–eval–print loop), with a searchable history, tab-completion, many helpful key-bindings, and dedicated help and shell modes; which can be used to experiment and test code quickly.The following fragment represents a sample session example where strings are concatenated automatically by println:

```
julia> p(x) = 2x^2 + 1; f(x, y) = 1 + 2p(x)y

julia> println("Hello world!", " I'm on cloud ", f(0, 4), " as Julia
supports recognizable syntax!")

Hello world! I'm on cloud 9 as Julia supports recognizable syntax!
```

The REPL gives user access to the system shell and to help mode, by pressing `;` or `?` after the prompt (preceding each command), respectively. It also keeps the history of commands, including between sessions. Code that can be tested inside the Julia's interactive section or saved into a file with a `.jl` extension and run from the command line by typing:

```
$ julia <filename>
```

Julia is supported by Jupyter, an online interactive "notebooks" environment.

Use with other Languages

Julia is in practice interoperable with many languages. Julia's *ccall* keyword is used to call C-exported or Fortran shared library functions individually.

Julia has support for the current Unicode 12.1 (which adds only one letter since Unicode 12.0), with UTF-8 used for strings (by default) and for Julia source code, meaning also allowing as an option common math symbols for many operators, such as ∈ for the `in` operator.

Julia has packages supporting markup languages such as HTML (and also for HTTP), XML, JSON and BSON, and for databases and web use in general.

Implementation

Julia's core is implemented in Julia and C, together with C++ for the LLVM dependency. The parsing and code-lowering are implemented in FemtoLisp, a Scheme dialect. The

LLVM compiler infrastructure project is used as the back end for generation of 64-bit or 32-bitoptimized machine code depending on the platform Julia runs on. With some exceptions (e.g., PCRE), the standard library is implemented in Julia itself. The most notable aspect of Julia's implementation is its speed, which is often within a factor of two relative to fully optimized C code (and thus often an order of magnitude faster than Python or R). Development of Julia began in 2009 and an open-source version was publicized in February 2012.

Current and Future Platforms

While Julia uses JIT, Julia generates native machine code directly, before a function is first run (not bytecodes that are run on a virtual machine (VM) or translated as the bytecode is running, as with, *e.g.*, Java; the JVM or Dalvik in Android).

Julia has four support tiers, and currently supports all x86-64 processors, that are 64-bit(and is more optimized for the latest generations) and all IA-32 ("x86") processors except for decades old ones, i.e., in 32-bit mode ("i686", excepting CPUs from the pre-Pentium 4-era); and supports more in lower tiers, e.g., ARM has tier 2 support: Julia "fully supports ARMv8 (AArch64) processors, and supports ARMv7 and ARMv6 (AArch32) with some caveats". CUDA (i.e. Nvidia GPUs; implementing PTX) has tier 1 support, with the help of an external package. There are also additionally packages supporting other accelerators, such as Google's TPUs, and AMD's GPUs also have support with e.g. OpenCL. Julia's downloads page provides executables (and source) for all the officially supported platforms.

On some platforms, Julia may need to be compiled from source code (e.g., the original Raspberry Pi), with specific build options. Julia has been "successfully built" on several ARM platforms, up to, e.g., "ARMv8 Data Center & Cloud Processors", such as Cavium ThunderX (first ARM with 48 cores). PowerPC (64-bit) has tier 3 support meaning "may or may not build". Julia is now supported in Raspbian while support is better for newer Pis, e.g. those with ARMv7 or newer; the Julia support is promoted by the Raspberry Pi Foundation. There's also support for web browsers/JavaScript through JSExpr.jl; and the alternative language of web browsers, WebAssembly, has support through minimal support in Julia itself, that supports several upcoming external Julia projects. Julia can compile to ARM; thus in theory Android apps can be made, but for now only Julia itself (and most packages) has been made to run under Android indirectly, i.e. in Ubuntu on Android.

LAPACK

LAPACK (Linear Algebra Package) is a standard software library for numerical linear algebra. It provides routines for solving systems of linear equations and linear least squares, eigenvalue problems, and singular value decomposition. It also includes routines to implement the associated matrix factorizations such as LU, QR, Cholesky and Schur decomposition. LAPACK was originally written in FORTRAN 77, but moved to Fortran 90 in version 3.2. The routines handle both real and complex matrices in both single and double precision.

LAPACK was designed as the successor to the linear equations and linear least-squares routines of LINPACK and the eigenvalue routines of EISPACK. LINPACK, written in the 1970s and 1980s, was designed to run on the then-modern vector computers with shared memory. LAPACK, in contrast, was designed to effectively exploit the caches on modern cache-based architectures, and thus can run orders of magnitude faster than LINPACK on such machines, given a well-tuned BLAS implementation. LAPACK has also been extended to run on distributed memory systems in later packages such as ScaLAPACK and PLAPACK.

LAPACK is licensed under a three-clause BSD style license, a permissive free software licensewith few restrictions.

Naming Scheme

Subroutines in LAPACK have a naming convention which makes the identifiers very compact. This was necessary as the first Fortran standards only supported identifiers up to six characters long, so the names had to be shortened to fit into this limit.

A LAPACK subroutine name is in the form *pmmaaa*, where:

- `p` is a one-letter code denoting the type of numerical constants used. `S`, `D` stand for real floating point arithmetic respectively in single and double precision, while `C` and `Z` stand for complex arithmetic with respectively single and double precision. The newer version, LAPACK95, uses generic subroutines in order to overcome the need to explicitly specify the data type.

- `mm` is a two-letter code denoting the kind of matrix expected by the algorithm. The codes for the different kind of matrices are reported below; the actual data are stored in a different format depending on the specific kind; e.g., when the code `DI` is given, the subroutine expects a vector of length `n` containing the elements on the diagonal, while when the code `GE` is given, the subroutine expects an $n \times n$ array containing the entries of the matrix.

- `aaa` is a one- to three-letter code describing the actual algorithm implemented in the subroutine, e.g. `SV` denotes a subroutine to solve linear system, while `R` denotes a rank-1 update.

For example, the subroutine to solve a linear system with a general (non-structured) matrix using real double-precision arithmetic is called DGESV.

Matrix Types in the LAPACK Naming Scheme	
Name	Description
BD	bidiagonal matrix
DI	diagonal matrix
GB	general band matrix
GE	general matrix (i.e., unsymmetric, in some cases rectangular)
GG	general matrices, generalized problem (i.e., a pair of general matrices)
GT	general tridiagonal matrix
HB	(complex) Hermitian band matrix
HE	(complex) Hermitian matrix
HG	upper Hessenberg matrix, generalized problem (i.e. a Hessenberg and a triangular matrix)
HP	(complex) Hermitian, packed storage matrix
HS	upper Hessenberg matrix
OP	(real) orthogonal matrix, packed storage matrix
OR	(real) orthogonal matrix
PB	symmetric matrix or Hermitian matrix positive definite band
PO	symmetric matrix or Hermitian matrix positive definite
PP	symmetric matrix or Hermitian matrix positive definite, packed storage matrix
PT	symmetric matrix or Hermitian matrix positive definite tridiagonal matrix
SB	(real) symmetric band matrix
SP	symmetric, packed storage matrix
ST	(real) symmetric matrix tridiagonal matrix
SY	symmetric matrix
TB	triangular band matrix
TG	triangular matrices, generalized problem (i.e., a pair of triangular matrices)
TP	triangular, packed storage matrix
TR	triangular matrix (or in some cases quasi-triangular)
TZ	trapezoidal matrix
UN	(complex) unitary matrix
UP	(complex) unitary, packed storage matrix

Use with other Programming Languages

Many programming environments today support the use of libraries with C binding. The LAPACK routines can be used like C functions if a few restrictions are observed.

Several alternative language bindings are also available:

- Armadillo for C++.
- IT++ for C++.
- LAPACK++ for C++.

- Lacaml for OCaml.
- CLapack for C.
- SciPy for Python.

LINEAR ALGEBRA LIBRARY

Lis (Library of Iterative Solvers for linear systems) is a scalable parallel software library for solving discretized linear equations and eigenvalue problems that mainly arise in the numerical solution of partial differential equations by using iterative methods. Although it is designed for parallel computers, the library can be used without being conscious of parallel processing.

Features

Lis provides facilities for:

- Automatic program configuration.
- NUMA aware hybrid implementation with MPI and OpenMP.
- Exchangeable dense and sparse matrix storage formats.
- Basic linear algebra operations for dense and sparse matrices.
- Parallel iterative methods for linear equations and eigenvalue problems.
- Parallel preconditioners for iterative methods.
- Quadruple precision floating point operations.
- Performance analysis.
- Command-line interface to solvers and benchmarks.

Example:

A C program to solve the linear equation $Ax = b$ is written as follows:

```
#include <stdio.h>

#include "lis_config.h"

#include "lis.h"
```

```
LIS_INT main(LIS_INT argc, char* argv[])
{
  LIS_MATRIX  A;
  LIS_VECTOR  b, x;
  LIS_SOLVER  solver;
  LIS_INT     iter;
  double      time;

  lis_initialize(&argc, &argv);

  lis_matrix_create(LIS_COMM_WORLD, &A);
  lis_vector_create(LIS_COMM_WORLD, &b);
  lis_vector_create(LIS_COMM_WORLD, &x);

  lis_input_matrix(A, argv);
  lis_input_vector(b, argv);
  lis_vector_duplicate(A, &x);

  lis_solver_create(&solver);
  lis_solver_set_optionC(solver);
  lis_solve(A, b, x, solver);

  lis_solver_get_iter(solver, &iter);
  lis_solver_get_time(solver, &time);
  printf("number of iterations = %d\n", iter);
  printf("elapsed time = %e\n", time);

  lis_output_vector(x, LIS_FMT_MM, argv[3]);
```

```
    lis_solver_destroy(solver);
    lis_matrix_destroy(A);
    lis_vector_destroy(b);
    lis_vector_destroy(x);

    lis_finalize();

    return 0;
}
```

System Requirements

The installation of Lis requires a C compiler. The Fortran interface requires a Fortran compiler, and the algebraic multigrid preconditioner requires a Fortran 90 compiler. For parallel computing environments, an OpenMP or MPI library is used. Both the Matrix Market and Harwell-Boeing formats are supported to import and export user data.

Packages that use Lis

- Gerris.
- OpenModelica.
- OpenGeoSys.
- SICOPOLIS.
- STOMP.
- Diablo.
- Kiva.
- Notus.
- Comet.

MATRIX-FREE METHODS

In computational mathematics, a matrix-free method is an algorithm for solving a linear system of equationsor an eigenvalue problem that does not store the coefficient matrix explicitly, but accesses the matrix by evaluating matrix-vector products. Such methods can be preferable when the matrix is so big that storing and manipulating it would cost a lot of memory and computer time, even with the use of methods for sparse matrices. Many iterative methods allow for a matrix-free implementation, including:

- The power method.
- The Lanczos algorithm.
- Locally Optimal Block Preconditioned Conjugate Gradient Method (LOBPCG).
- Wiedemann's coordinate recurrence algorithm.
- The conjugate gradient method.

Distributed solutions have also been explored using coarse-grain parallel software systems to achieve homogeneous solutions of linear systems.

It is generally used in solving non-linear equations like Euler's equations in Computational Fluid Dynamics. Solving these equations requires the calculation of the jacobian which is costly in terms of CPU time and storage. To avoid this expense, matrix free methods are employed. In order to remove the need to calculate the jacobian, the jacobian vector product is formed instead, which is in fact a vector itself. Manipulating and calculating this vector is easier than working with a large matrix or linear system.

MINIMUM DEGREE ALGORITHM

In numerical analysis the minimum degree algorithm is an algorithm used to permute the rows and columns of a symmetric sparse matrix before applying the Cholesky decomposition, to reduce the number of non-zeros in the Cholesky factor. This results in reduced storage requirements and means that the Cholesky factor can be applied with fewer arithmetic operations. Sometimes it may also pertain to an incomplete Cholesky factor used as a preconditioner, for example in the preconditioned conjugate gradient algorithm.

Minimum degree algorithms are often used in the finite element method where the reordering of nodes can be carried out depending only on the topology of the mesh, rather than the coefficients in the partial differential equation, resulting in efficiency savings when the same mesh is used for a variety of coefficient values.

Given a linear system:

$$Ax=b$$

where A is an $n \times n$ real symmetric sparse square matrix the Cholesky factor L will typically suffer 'fill in', that is have more non-zeros than the upper triangle of A. We seek a permutation matrix P, so that the matrix $P^T AP$, which is also symmetric, has the least possible fill in its Cholesky factor. We solve the reordered system:

$$\left(P^T AP\right)\left(P^T x\right) = P^T b.$$

The problem of finding the best ordering is an NP-complete problem and is thus intractable, so heuristic methods are used instead. The minimum degree algorithm is derived from a method first proposed by Markowitz in 1959 for non-symmetric linear programming problems, which is loosely described as follows. At each step in Gaussian elimination row and column permutations are performed so as to minimize the number of off diagonal non-zeros in the pivot row and column. A symmetric version of Markowitz method was described by Tinney and Walker in 1967 and Rose later derived a graph theoretic version of the algorithm where the factorization is only simulated, and this was named the minimum degree algorithm. The graph referred to is the graph with n vertices, with vertices i and j connected by an edge when $a_{ij} \neq 0$, and the *degree* is the degree of the vertices. A crucial aspect of such algorithms is a tie breaking strategy when there is a choice of renumbering resulting in the same degree.

A version of the minimum degree algorithm was implemented in the MATLAB function symmmd (where MMD stands for multiple minimum degree), but has now been superseded by a symmetric approximate multiple minimum degree function symamd, which is faster. This is confirmed by theoretical analysis, which shows that for graphs on n vertices and m edges, MMD has a tight upper bound of $O(n^2m)$ on its run time, whereas for AMD a tight bound of $O(nm)$ holds.

MATLAB

MATLAB is a high-performance language for technical computing. It integrates computation, visualization, and programming in an easy-to-use environment where problems and solutions are expressed in familiar mathematical notation. Typical uses include:

- Math and computation.
- Algorithm development.
- Modeling, simulation, and prototyping.
- Data analysis, exploration, and visualization.
- Scientific and engineering graphics.
- Application development, including Graphical User Interface building.

MATLAB is an interactive system whose basic data element is an array that does not require dimensioning. This allows you to solve many technical computing problems, especially those with matrix and vector formulations, in a fraction of the time it would take to write a program in a scalar noninteractive language such as C or Fortran.

The name MATLAB stands for matrix laboratory. MATLAB was originally written to provide easy access to matrix software developed by the LINPACK and EISPACK projects, which together represent the state-of-the-art in software for matrix computation.

MATLAB has evolved over a period of years with input from many users. In university environments, it is the standard instructional tool for introductory and advanced courses in mathematics, engineering, and science. In industry, MATLAB is the tool of choice for high-productivity research, development, and analysis.

MATLAB features a family of application-specific solutions called toolboxes. Very important to most users of MATLAB, toolboxes allow you to *learn* and *apply* specialized technology. Toolboxes are comprehensive collections of MATLAB functions (M-files) that extend the MATLAB environment to solve particular classes of problems. Areas in which toolboxes are available include signal processing, control systems, neural networks, fuzzy logic, wavelets, simulation, and many others.

MATLAB System

The MATLAB system consists of five main parts:

MATLAB Language

This is a high-level matrix/array language with control flow statements, functions, data structures, input/output, and object-oriented programming features. It allows both "programming in the small" to rapidly create quick and dirty throw-away programs, and "programming in the large" to create complete large and complex application programs.

MATLAB Working Environment

This is the set of tools and facilities that you work with as the MATLAB user or programmer. It includes facilities for managing the variables in your workspace and importing and exporting data. It also includes tools for developing, managing, debugging, and profiling M-files, MATLAB's applications.

Handle Graphics

This is the MATLAB graphics system. It includes high-level commands for two-dimensional and three-dimensional data visualization, image processing, animation, and presentation graphics. It also includes low-level commands that allow you to fully customize the appearance of graphics as well as to build complete Graphical User Interfaces on your MATLAB applications.

MATLAB Mathematical Function Library

This is a vast collection of computational algorithms ranging from elementary functions like sum, sine, cosine, and complex arithmetic, to more sophisticated functions like matrix inverse, matrix eigenvalues, Bessel functions, and fast Fourier transforms.

MATLAB Application Program Interface (API)

This is a library that allows you to write C and Fortran programs that interact with MATLAB. It include facilities for calling routines from MATLAB (dynamic linking), calling MATLAB as a computational engine, and for reading and writing MAT-files.

References

- Whatismatlab: cimss.ssec.wisc.edu, Retrieved 28 August, 2019
- Conrad Sanderson and Ryan Curtin (2016). "Armadillo: a template-based C++ library for linear algebra". Journal of Open Source Software. 1 (2): 26. Bibcode:2016JOSS....1...26S. doi:10.21105/joss.00026
- Simoncini, V. (2016). "Computational Methods for Linear Matrix Equations". SIAM Review. 58 (3): 377–441. doi:10.1137/130912839. ISSN 0036-1445
- R. Clint Whaley; Antoine Petitet & Jack J. Dongarra (2001). "Automated Empirical Optimization of Software and the ATLAS Project" (PDF). Parallel Computing. 27 (1–2): 3–35. CiteSeerX 10.1.1.35.2297. doi:10.1016/S0167-8191(00)00087-9. Retrieved 2006-10-06
- Goto, Kazushige; van de Geijn, Robert A. (2008). "Anatomy of High-Performance Matrix Multiplication". ACM Transactions on Mathematical Software. 34 (3): Article 12, 25 pages. CiteSeerX 10.1.1.111.3873. doi:10.1145/1356052.1356053

Mathematical Optimization

5

CHAPTER

Mathematical optimization refers to a collection of mathematical principles and methods to solve optimization problems. Linear and non-linear programming, and dynamic programming are also studied under it. This chapter closely examines the aspects related to mathematical optimization for an in-depth understanding of the subject.

Optimization is also known as mathematical programming is a collection of mathematical. Principles and methods used for solving quantitative problems in many disciplines, including physics, biology, engineering, economics, and business. The subject grew from a realization that quantitative problems in manifestly different disciplines have important mathematical elements in common. Because of this commonality, many problems can be formulated and solved by using the unified set of ideas and methods that make up the field of optimization.

The historic term mathematical programming, broadly synonymous with optimization, was coined in the 1940s before programming became equated with computer programming. Mathematical programming includes the study of the mathematical structure of optimization problems, the invention of methods for solving these problems, the study of the mathematical properties of these methods, and the implementation of these methods on computers. Faster computers have greatly expanded the size and complexity of optimization problems that can be solved. The development of optimization techniques has paralleled advances not only in computer science but also in operations research, numerical analysis, game theory, mathematical economics, control theory, and combinatorics.

Optimization problems typically have three fundamental elements. The first is a single numerical quantity, or objective function, that is to be maximized or minimized. The objective may be the expected return on a stock portfolio, a company's production costs or profits, the time of arrival of a vehicle at a specified destination, or the vote share of a political candidate. The second element is a collection of variables, which are quantities whose values can be manipulated in order to optimize the objective. Examples include the quantities of stock to be bought or sold, the amounts of various resources to be allocated to different production activities, the route to be followed by a vehicle through a traffic network, or the policies to be advocated by a candidate. The third element of an optimization problem is a set of constraints, which are restrictions on the values that the variables can take. For instance, a manufacturing process cannot require more resources than are available, nor can it employ less than zero resources. Within this broad framework, optimization problems can have different mathematical properties. Problems in which the variables are continuous quantities (as in the resource allocation example) require a different approach from problems in which the variables are discrete or combinatorial quantities (as in the selection of a vehicle route from among a predefined set of possibilities).

An important class of optimization is known as linear programming. Linear indicates that no variables are raised to higher powers, such as squares. For this class, the problems involve minimizing (or maximizing) a linear objective function whose variables are real numbers that are constrained to satisfy a system of linear equalities and inequalities. Another important class of optimization is known as nonlinear programming. In nonlinear programming the variables are real numbers, and the objective or some of the constraints are nonlinear functions (possibly involving squares, square roots, trigonometric functions, or products of the variables). Other important classes of optimization problems not covered here includes stochastic programming, in which the objective function or the constraints depend on random variables, so that the optimum is found in some "expected," or probabilistic, sense; network optimization, which involves optimization of some property of a flow through a network, such as the maximization of the amount of material that can be transported between two given locations in the network; and combinatorial optimization, in which the solution must be found among a finite but very large set of possible values, such as the many possible ways to assign 20 manufacturing plants to 20 locations.

PROGRAM OPTIMIZATION

In computer science, program optimization or software optimization is the process of modifying a software system to make some aspect of it work more efficiently or use fewer resources. In general, a computer program may be optimized so that it executes more rapidly, or to make it capable of operating with less memory storage or other resources, or draw less power.

Although the word "optimization" shares the same root as "optimal", it is rare for the process of optimization to produce a truly optimal system. A system can generally be made optimal not in absolute terms, but only with respect to a given quality metrics, that may be in contrast with other possible metrics. As a result, the optimized system will typically only be optimal in one application or for one audience. One might reduce the amount of time that a program takes to perform some task at the price of making it consume more memory. In an application where memory space is at a premium, one might deliberately choose a slower algorithm in order to use less memory. Often there is no "one size fits all" design which works well in all cases, so engineers make trade-offs to optimize the attributes of greatest interest. Additionally, the effort required to make a piece of software completely optimal — incapable of any further improvement — is almost always more than is reasonable for the benefits that would be accrued; so the process of optimization may be halted before a completely optimal solution has been reached. Fortunately, it is often the case that the greatest improvements come early in the process.

Levels of Optimization

Optimization can occur at a number of levels. Typically the higher levels have greater impact, and are harder to change later on in a project, requiring significant changes or a complete rewrite if they need to be changed. Thus optimization can typically proceed via refinement from higher to lower, with initial gains being larger and achieved with less work, and later gains being smaller and requiring more work. However, in some cases overall performance depends on performance of very low-level portions of a program, and small changes at a late stage or early consideration of low-level details can have outsized impact. Typically some consideration is given to efficiency throughout a project – though this varies significantly – but major optimization is often considered a refinement to be done late, if ever. On longer-running projects there are typically cycles of optimization, where improving one area reveals limitations in another, and these are typically curtailed when performance is acceptable or gains become too small or costly.

As performance is part of the specification of a program – a program that is unusably slow is not fit for purpose: a video game with 60 Hz (frames-per-second) is acceptable, but 6 frames-per-second is unacceptably choppy – performance is a consideration from the start, to ensure that the system is able to deliver sufficient performance, and early prototypes need to have roughly acceptable performance for there to be confidence that the final system will (with optimization) achieve acceptable performance. This is sometimes omitted in the belief that optimization can always be done later, resulting in prototype systems that are far too slow – often by an order of magnitude or more – and systems that ultimately are failures because they architecturally cannot achieve their performance goals, such as the Intel 432; or ones that take years of work to achieve acceptable performance, such as Java, which only achieved acceptable performance with HotSpot. The degree to which performance changes between prototype and production system, and how amenable it is to optimization, can be a significant source of uncertainty and risk.

Design Level

At the highest level, the design may be optimized to make best use of the available resources, given goals, constraints, and expected use/load. The architectural design of a system overwhelmingly affects its performance. For example, a system that is network latency-bound (where network latency is the main constraint on overall performance) would be optimized to minimize network trips, ideally making a single request (or no requests, as in a push protocol) rather than multiple roundtrips. Choice of design depends on the goals: when designing a compiler, if fast compilation is the key priority, a one-pass compiler is faster than a multi-pass compiler (assuming same work), but if speed of output code is the goal, a slower multi-pass compiler fulfills the goal better, even though it takes longer itself. Choice of platform and programming language occur at this level, and changing them frequently requires a complete rewrite, though a modular system may allow rewrite of only some component – for example, a Python program may rewrite performance-critical sections in C. In a distributed system, choice of architecture (client-server, peer-to-peer, etc.) occurs at the design level, and may be difficult to change, particularly if all components cannot be replaced in sync (e.g., old clients).

Algorithms and Data Structures

Given an overall design, a good choice of efficient algorithms and data structures, and efficient implementation of these algorithms and data structures comes next. After design, the choice of algorithms and data structures affects efficiency more than any other aspect of the program. Generally data structures are more difficult to change than algorithms, as a data structure assumption and its performance assumptions are used throughout the program, though this can be minimized by the use of abstract data types in function definitions, and keeping the concrete data structure definitions restricted to a few places.

For algorithms, this primarily consists of ensuring that algorithms are constant $O(1)$, logarithmic $O(\log n)$, linear $O(n)$, or in some cases log-linear $O(n \log n)$ in the input (both in space and time). Algorithms with quadratic complexity $O(n^2)$ fail to scale, and even linear algorithms cause problems if repeatedly called, and are typically replaced with constant or logarithmic if possible.

Beyond asymptotic order of growth, the constant factors matter: an asymptotically slower algorithm may be faster or smaller (because simpler) than an asymptotically faster algorithm when they are both faced with small input, which may be the case that occurs in reality. Often a hybrid algorithm will provide the best performance, due to this tradeoff changing with size.

A general technique to improve performance is to avoid work. A good example is the use of a fast path for common cases, improving performance by avoiding unnecessary work. For example, using a simple text layout algorithm for Latin text, only switching to a complex layout algorithm for complex scripts, such as Devanagari. Another important technique is caching, particularly memoization, which avoids redundant computations. Because of the importance of caching, there are often many levels of caching in a system, which can cause problems from memory use, and correctness issues from stale caches.

Source Code Level

Beyond general algorithms and their implementation on an abstract machine, concrete source code level choices can make a significant difference. For example, on early C compilers, while(1) was slower than for(;;) for an unconditional loop, because while(1) evaluated 1 and then had a conditional jump which tested if it was true, while for (;;) had an unconditional jump. Some optimizations (such as this one) can nowadays be performed by optimizing compilers. This depends on the source language, the target machine language, and the compiler, and can be both difficult to understand or predict and changes over time; this is a key place where understanding of compilers and machine code can improve performance. Loop-invariant code motion and return value optimization are examples of optimizations that reduce the need for auxiliary variables and can even result in faster performance by avoiding round-about optimizations.

Build Level

Between the source and compile level, directives and build flags can be used to tune performance options in the source code and compiler respectively, such as using preprocessor defines to disable unneeded software features, optimizing for specific processor models or hardware capabilities, or predicting branching, for instance. Source-based software distribution systems such as BSD's Ports and Gentoo's Portage can take advantage of this form of optimization.

Compile Level

Use of an optimizing compiler tends to ensure that the executable program is optimized at least as much as the compiler can predict.

Assembly Level

At the lowest level, writing code using an assembly language, designed for a particular hardware platform can produce the most efficient and compact code if the programmer takes advantage of the full repertoire of machine instructions. Many operating systems used on embedded systems have been traditionally written in assembler code for this reason. Programs (other than very small programs) are seldom written from start to finish in assembly due to the time and cost involved. Most are compiled down from a high level language to assembly and hand optimized from there. When efficiency and size are less important large parts may be written in a high-level language.

With more modern optimizing compilers and the greater complexity of recent CPUs, it is harder to write more efficient code than what the compiler generates, and few projects need this "ultimate" optimization step.

Much code written today is intended to run on as many machines as possible. As a consequence, programmers and compilers don't always take advantage of the more efficient instructions provided by newer CPUs or quirks of older models. Additionally, assembly code tuned for a particular processor without using such instructions might still be suboptimal on a different processor, expecting a different tuning of the code.

Typically today rather than writing in assembly language, programmers will use a disassembler to analyze the output of a compiler and change the high-level source code so that it can be compiled more efficiently, or understand why it is inefficient.

Run Time

Just-in-time compilers can produce customized machine code based on run-time data, at the cost of compilation overhead. This technique dates to the earliest regular expression engines, and has become widespread with Java HotSpot and V8 for JavaScript. In some cases adaptive optimization may be able to perform run time optimization exceeding the capability of static compilers by dynamically adjusting parameters according to the actual input or other factors.

Profile-guided optimization is an ahead-of-time (AOT) compilation optimization technique based on run time profiles, and is similar to a static "average case" analog of the dynamic technique of adaptive optimization.

Self-modifying code can alter itself in response to run time conditions in order to optimize code; this was more common in assembly language programs.

Some CPU designs can perform some optimizations at run time. Some examples include Out-of-order execution, Speculative execution, Instruction pipelines, and Branch predictors. Compilers can help the program take advantage of these CPU features, for example through instruction scheduling.

Platform Dependent and Independent Optimizations

Code optimization can be also broadly categorized as platform-dependent and platform-independent techniques. While the latter ones are effective on most or all platforms, platform-dependent techniques use specific properties of one platform, or rely on parameters depending on the single platform or even on the single processor. Writing or producing different versions of the same code for different processors might therefore be needed. For instance, in the case of compile-level optimization, platform-independent techniques are generic techniques (such as loop unrolling, reduction in function calls, memory efficient routines, reduction in conditions, etc.), that impact most CPU architectures in a similar way. A great example of platform-independent optimization has been shown with inner for loop, where it was observed that a loop with an inner for loop performs more computations per unit time than a loop without it or one with an inner while loop. Generally, these serve to reduce the total instruction path length required to complete the program and reduce total memory usage during the process. On the other hand, platform-dependent techniques involve instruction scheduling, instruction-level parallelism, data-level parallelism, cache optimization techniques (i.e., parameters that differ among various platforms) and the optimal instruction scheduling might be different even on different processors of the same architecture.

Strength Reduction

Computational tasks can be performed in several different ways with varying efficiency. A more efficient version with equivalent functionality is known as a strength reduction. For example, consider the following C code snippet whose intention is to obtain the sum of all integers from 1 to N:

```
int i, sum = 0;

for (i = 1; i <= N; ++i) {

  sum += i;

}

printf("sum: %d\n", sum);
```

This code can (assuming no arithmetic overflow) be rewritten using a mathematical formula like:

```
int sum = N * (1 + N) / 2;

printf("sum: %d\n", sum);
```

The optimization, sometimes performed automatically by an optimizing compiler, is to select a method (algorithm) that is more computationally efficient, while retaining the same functionality.

Optimization is not always an obvious or intuitive process. In the example above, the "optimized" version might actually be slower than the original version if N were sufficiently small and the particular hardware happens to be much faster at performing addition and looping operations than multiplication and division.

Trade-offs

In some cases, however, optimization relies on using more elaborate algorithms, making use of "special cases" and special "tricks" and performing complex trade-offs. A "fully optimized" program might be more difficult to comprehend and hence may contain more faults than unoptimized versions. Beyond eliminating obvious antipatterns, some code level optimizations decrease maintainability.

Optimization will generally focus on improving just one or two aspects of performance: execution time, memory usage, disk space, bandwidth, power consumption or some other resource. This will usually require a trade-off — where one factor is optimized at the expense of others. For example, increasing the size of cache improves run time performance, but also increases the memory consumption. Other common trade-offs include code clarity and conciseness.

There are instances where the programmer performing the optimization must decide to make the software better for some operations but at the cost of making other operations less efficient. These trade-offs may sometimes be of a non-technical nature – such as when a competitor has published a benchmark result that must be beaten in order to improve commercial success but comes perhaps with the burden of making normal usage of the software less efficient. Such changes are sometimes jokingly referred to as *pessimizations*.

Bottlenecks

Optimization may include finding a bottleneck in a system – a component that is the limiting factor on performance. In terms of code, this will often be a hot spot – a critical part of the code that is the primary consumer of the needed resource – though it can be another factor, such as I/O latency or network bandwidth.

In computer science, resource consumption often follows a form of power law distribution, and the Pareto principle can be applied to resource optimization by observing that 80% of the resources are typically used by 20% of the operations. In software engineering, it is often a better approximation that 90% of the execution time of a computer program is spent executing 10% of the code (known as the 90/10 law in this context).

More complex algorithms and data structures perform well with many items, while simple algorithms are more suitable for small amounts of data – the setup, initialization time, and constant factors of the more complex algorithm can outweigh the benefit, and thus a hybrid algorithm or adaptive algorithm may be faster than any single algorithm. A performance profiler can be used to narrow down decisions about which functionality fits which conditions.

In some cases, adding more memory can help to make a program run faster. For example, a filtering program will commonly read each line and filter and output that line immediately. This only uses enough memory for one line, but performance is typically poor, due to the latency of each disk read. Caching the result is similarly effective, though also requiring larger memory use.

When to Optimize

Optimization can reduce readability and add code that is used only to improve the performance. This may complicate programs or systems, making them harder to maintain and debug. As a result, optimization or performance tuning is often performed at the end of the development stage.

Donald Knuth made the following two statements on optimization:

> "We should forget about small efficiencies, say about 97% of the time: premature optimization is the root of all evil. Yet we should not pass up our opportunities in that critical 3%".

He also attributed the quote to Tony Hoare several years later, although this might have been an error as Hoare disclaims having coined the phrase.

> "In established engineering disciplines a 12% improvement, easily obtained, is never considered marginal and I believe the same viewpoint should prevail in software engineering".

"Premature optimization" is a phrase used to describe a situation where a programmer lets performance considerations affect the design of a piece of code. This can result in a design that is not as clean as it could have been or code that is incorrect, because the code is complicated by the optimization and the programmer is distracted by optimizing.

When deciding whether to optimize a specific part of the program, Amdahl's Law should always be considered: the impact on the overall program depends very much on how much time is actually spent in that specific part, which is not always clear from looking at the code without a performance analysis.

A better approach is therefore to design first, code from the design and then profile/benchmark the resulting code to see which parts should be optimized. A simple and elegant design is often easier to optimize at this stage, and profiling may reveal unexpected performance problems that would not have been addressed by premature optimization.

In practice, it is often necessary to keep performance goals in mind when first designing software, but the programmer balances the goals of design and optimization.

Modern compilers and operating systems are so efficient that the intended performance increases often fail to materialize. As an example, caching data at the application level that is again cached at the operating system level does not yield improvements in execution. Even so, it is a rare case when the programmer will remove failed optimizations from production code. It is also true that advances in hardware will more often than not obviate any potential improvements, yet the obscuring code will persist into the future long after its purpose has been negated.

Macros

Optimization during code development using macros takes on different forms in different languages.

In some procedural languages, such as C and C++, macros are implemented using token substitution. Nowadays, inline functions can be used as a type safe alternative in many cases. In both cases, the inlined function body can then undergo further compile-time optimizations by the compiler, including constant folding, which may move some computations to compile time.

In many functional programming languages macros are implemented using parse-time substitution of parse trees/abstract syntax trees, which it is claimed makes them safer to use. Since in many cases interpretation is used, that is one way to ensure that such computations are only performed at parse-time, and sometimes the only way.

Lisp originated this style of macro and such macros are often called "Lisp-like macros." A similar effect can be achieved by using template metaprogramming in C++.

In both cases, work is moved to compile-time. The difference between C macros on one side, and Lisp-like macros and C++ template metaprogramming on the other side, is that the latter tools allow performing arbitrary computations at compile-time/parse-time, while expansion of C macros does not perform any computation, and relies on the optimizer ability to perform it. Additionally, C macros do not directly support recursion or iteration, so are not Turing complete.

As with any optimization, however, it is often difficult to predict where such tools will have the most impact before a project is complete.

Automated and Manual Optimization

Optimization can be automated by compilers or performed by programmers. Gains are usually limited for local optimization, and larger for global optimizations. Usually, the most powerful optimization is to find a superior algorithm.

Optimizing a whole system is usually undertaken by programmers because it is too complex for automated optimizers. In this situation, programmers or system administrators explicitly change code so that the overall system performs better. Although it can produce better efficiency, it is far more expensive than automated optimizations. Since many parameters influence the program performance, the program optimization space is large. Meta-heuristics and machine learning are used to address the complexity of program optimization.

Use a profiler (or performance analyzer) to find the sections of the program that are taking the most resources — the *bottleneck*. Programmers sometimes believe they have a clear idea of where the bottleneck is, but intuition is frequently wrong. Optimizing an unimportant piece of code will typically do little to help the overall performance.

When the bottleneck is localized, optimization usually starts with a rethinking of the algorithm used in the program. More often than not, a particular algorithm can be specifically tailored to a particular problem, yielding better performance than a generic algorithm. For example, the task of sorting a huge list of items is usually done with a quicksort routine, which is one of the most efficient generic algorithms. But if some characteristic of the items is exploitable (for example, they are already arranged in some particular order), a different method can be used, or even a custom-made sort routine.

After the programmer is reasonably sure that the best algorithm is selected, code optimization can start. Loops can be unrolled (for lower loop overhead, although this can often lead to *lower* speed if it overloads the CPU cache), data types as small as possible can be used, integer arithmetic can be used instead of floating-point, and so on.

Performance bottlenecks can be due to language limitations rather than algorithms or data structures used in the program. Sometimes, a critical part of the program can be re-written in a different programming language that gives more direct access to the underlying machine. For example, it is common for very high-level languages like Python to have modules written in C for greater speed. Programs already written in C can have modules written in assembly. Programs written in D can use the inline assembler.

Rewriting sections "pays off" in these circumstances because of a general "rule of thumb" known as the 90/10 law, which states that 90% of the time is spent in 10% of the code, and only 10% of the time in the remaining 90% of the code. So, putting intellectual effort into optimizing just a small part of the program can have a huge effect on the overall speed — if the correct part(s) can be located.

Manual optimization sometimes has the side effect of undermining readability. Thus code optimizations should be carefully documented (preferably using in-line comments), and their effect on future development evaluated.

The program that performs an automated optimization is called an optimizer. Most optimizers are embedded in compilers and operate during compilation. Optimizers can often tailor the generated code to specific processors.

Today, automated optimizations are almost exclusively limited to compiler optimization. However, because compiler optimizations are usually limited to a fixed set of rather general optimizations, there is considerable demand for optimizers which can accept descriptions of problem and language-specific optimizations, allowing an engineer to specify custom optimizations. Tools that accept descriptions of optimizations are called program transformation systems and are beginning to be applied to real software systems such as C++.

Some high-level languages (Eiffel, Esterel) optimize their programs by using an intermediate language.

Grid computing or distributed computing aims to optimize the whole system, by moving tasks from computers with high usage to computers with idle time.

Time taken for Optimization

Sometimes, the time taken to undertake optimization therein itself may be an issue.

Optimizing existing code usually does not add new features, and worse, it might add new bugs in previously working code (as any change might). Because manually optimized code might sometimes have less "readability" than unoptimized code, optimization might impact maintainability of it as well. Optimization comes at a price and it is important to be sure that the investment is worthwhile.

An automatic optimizer (or optimizing compiler, a program that performs code optimization) may itself have to be optimized, either to further improve the efficiency of its target programs or else speed up its own operation. A compilation performed with optimization "turned on" usually takes longer, although this is usually only a problem when programs are quite large.

In particular, for just-in-time compilers the performance of the run time compile component, executing together with its target code, is the key to improving overall execution speed.

LOCAL OPTIMIZATION

In computer science, local search is a heuristic method for solving computationally hard optimization problems. Local search can be used on problems that can be formulated as finding a solution maximizing a criterion among a number of candidate solutions. Local search algorithms move from solution to solution in the space of candidate solutions (the *search space*) by applying local changes, until a solution deemed optimal is found or a time bound is elapsed.

Local search algorithms are widely applied to numerous hard computational problems, including problems from computer science (particularly artificial intelligence), mathematics, operations research, engineering, and bioinformatics. Examples of local search algorithms are WalkSAT, the 2-opt algorithm for the Traveling Salesman Problem and the Metropolis–Hastings algorithm.

Examples:

Some problems where local search has been applied are:

1. The vertex cover problem, in which a solution is a vertex cover of a graph, and the target is to find a solution with a minimal number of nodes.
2. The travelling salesman problem, in which a solution is a cycle containing all nodes of the graph and the target is to minimize the total length of the cycle.
3. The boolean satisfiability problem, in which a candidate solution is a truth assignment, and the target is to maximize the number of clauses satisfied by the assignment; in this case, the final solution is of use only if it satisfies *all* clauses.
4. The nurse scheduling problem where a solution is an assignment of nurses to shifts which satisfies all established constraints.

5. The k-medoid clustering problem and other related facility location problems for which local search offers the best known approximation ratios from a worst-case perspective.
6. The Hopfield Neural Networks problem for which finding stable configurations in Hopfield network.

Most problems can be formulated in terms of search space and target in several different manners. For example, for the travelling salesman problem a solution can be a cycle and the criterion to maximize is a combination of the number of nodes and the length of the cycle. But a solution can also be a path, and being a cycle is part of the target.

A local search algorithm starts from a candidate solution and then iteratively moves to a neighbor solution. This is only possible if a neighborhood relation is defined on the search space. As an example, the neighborhood of a vertex cover is another vertex cover only differing by one node. For boolean satisfiability, the neighbors of a truth assignment are usually the truth assignments only differing from it by the evaluation of a variable. The same problem may have multiple different neighborhoods defined on it; local optimization with neighborhoods that involve changing up to *k* components of the solution is often referred to as k-opt.

Typically, every candidate solution has more than one neighbor solution; the choice of which one to move to is taken using only information about the solutions in the neighborhood of the current one, hence the name *local* search. When the choice of the neighbor solution is done by taking the one locally maximizing the criterion, the metaheuristic takes the name hill climbing. When no improving configurations are present in the neighborhood, local search is stuck at a locally optimal point. This local-optima problem can be cured by using restarts (repeated local search with different initial conditions), or more complex schemes based on iterations, like iterated local search, on memory, like reactive search optimization, on memory-less stochastic modifications, like simulated annealing.

Termination of local search can be based on a time bound. Another common choice is to terminate when the best solution found by the algorithm has not been improved in a given number of steps. Local search is an anytime algorithm: it can return a valid solution even if it's interrupted at any time before it ends. Local search algorithms are typically approximation or incomplete algorithms, as the search may stop even if the best solution found by the algorithm is not optimal. This can happen even if termination is due to the impossibility of improving the solution, as the optimal solution can lie far from the neighborhood of the solutions crossed by the algorithms.

For specific problems it is possible to devise neighborhoods which are very large, possibly exponentially sized. If the best solution within the neighborhood can be found efficiently, such algorithms are referred to as very large-scale neighborhood search algorithms.

LINEAR PROGRAMMING

Linear programming (LP, also called linear optimization) is a method to achieve the best outcome (such as maximum profit or lowest cost) in a mathematical model whose requirements are represented by linear relationships. Linear programming is a special case of mathematical programming (also known as mathematical optimization).

More formally, linear programming is a technique for the optimization of a linear objective function, subject to linear equality and linear inequality constraints. Its feasible region is a convex polytope, which is a set defined as the intersection of finitely many half spaces, each of which is defined by a linear inequality. Its objective function is a real-valued affine (linear) function defined on this polyhedron. A linear programming algorithm finds a point in the polytope where this function has the smallest (or largest) value if such a point exists.

Linear programs are problems that can be expressed in canonical form as:

$$\begin{aligned} &\text{Maximize} && \mathbf{c}^{\mathrm{T}}\mathbf{x} \\ &\text{subject to} && A\mathbf{x} \leq \mathbf{b} \\ &\text{and} && \mathbf{x} \geq \mathbf{0} \end{aligned}$$

where x represents the vector of variables (to be determined), c and b are vectors of (known) coefficients, A is a (known) matrix of coefficients, and $(\cdot)^T$ is the matrix transpose. The expression to be maximized or minimized is called the objective function ($c^T x$ in this case). The inequalities $Ax \leq b$ and $x \geq 0$ are the constraints which specify a convex polytope over which the objective function is to be optimized. In this context, two vectors are comparable when they have the same dimensions. If every entry in the first is less-than or equal-to the corresponding entry in the second, then it can be said that the first vector is less-than or equal-to the second vector.

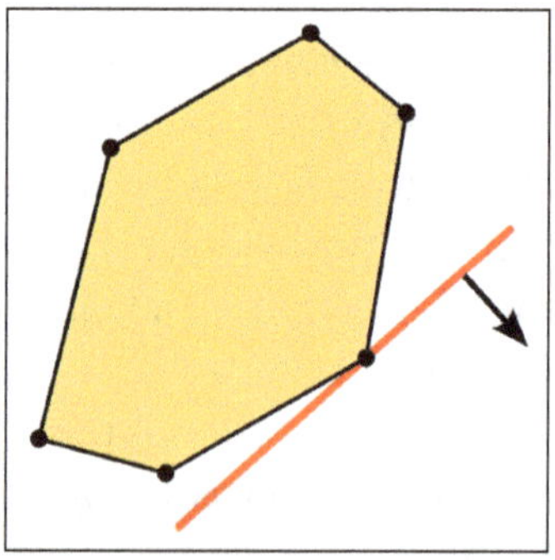

A pictorial representation of a simple linear program with two variables and six inequalities.

Linear programming can be applied to various fields of study. It is widely used in mathematics, and to a lesser extent in business, economics, and for some engineering problems. Industries that use linear programming models include transportation, energy, telecommunications, and manufacturing. It has proven useful in modeling diverse types of problems in planning, routing, scheduling, assignment, and design.

The set of feasible solutions is depicted in yellow and forms a polygon, a 2-dimensional polytope. The linear cost function is represented by the red line and the arrow: The red line is a level set of the cost function, and the arrow indicates the direction in which we are optimizing.

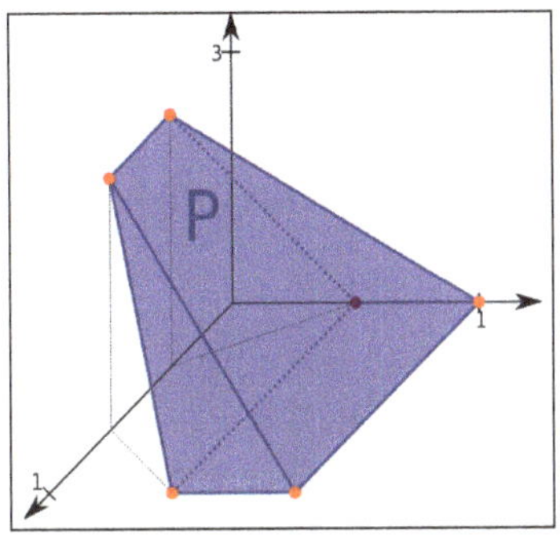

A closed feasible region of a problem with three variables is a convex polyhedron. The surfaces giving a fixed value of the objective function are planes. The linear programming problem is to find a point on the polyhedron that is on the plane with the highest possible value.

Uses

Linear programming is a widely used field of optimization for several reasons. Many practical problems in operations research can be expressed as linear programming problems. Certain special cases of linear programming, such as *network flow* problems and *multicommodity flow* problems are considered important enough to have generated much research on specialized algorithms for their solution. A number of algorithms for other types of optimization problems work by solving LP problems as sub-problems. Historically, ideas from linear programming have inspired many of the central concepts of optimization theory, such as *duality, decomposition,* and the importance of *convexity* and its generalizations. Likewise, linear programming was heavily used in the early formation of microeconomics and it is currently utilized in company management, such as planning, production, transportation, technology and other issues. Although the modern management issues are ever-changing, most companies would like to maximize profits and minimize costs with limited resources. Therefore, many issues can be characterized as linear programming problems.

Standard Form

Standard form is the usual and most intuitive form of describing a linear programming problem. It consists of the following three parts:

- A linear function to be maximized:

  e.g. $f(x_1,x_2)=c_1x_1+c_2x_2$

- Problem constraints of the following form:

$$\text{e.g.}\quad \begin{aligned} a_{11}x_1 + a_{12}x_2 &\le b_1 \\ a_{21}x_1 + a_{22}x_2 &\le b_2 \\ a_{31}x_1 + a_{32}x_2 &\le b_3 \end{aligned}$$

- Non-negative variables:

$$\text{e.g.}\quad \begin{aligned} x_1 &\ge 0 \\ x_2 &\ge 0 \end{aligned}$$

The problem is usually expressed in *matrix form*, and then becomes:

$$\max\{\mathrm{c}^T\mathrm{x} \mid A\mathrm{x} \le \mathrm{b} \wedge \mathrm{x} \ge 0\}$$

Other forms, such as minimization problems, problems with constraints on alternative forms, as well as problems involving negative variables can always be rewritten into an equivalent problem in standard form.

Example:

Suppose that a farmer has a piece of farm land, say L km², to be planted with either wheat or barley or some combination of the two. The farmer has a limited amount of fertilizer, F kilograms, and pesticide, P kilograms. Every square kilometer of wheat requires F_1 kilograms of fertilizer and P_1 kilograms of pesticide, while every square kilometer of barley requires F_2 kilograms of fertilizer and P_2 kilograms of pesticide. Let S_1 be the selling price of wheat per square kilometer, and S_2 be the selling price of barley. If we denote the area of land planted with wheat and barley by x_1 and x_2 respectively, then profit can be maximized by choosing optimal values for x_1 and x_2. This problem can be expressed with the following linear programming problem in the standard form:

Maximize: $S_1 \cdot x_1 + S_2 \cdot x_2$		(Maximize the revenue—revenue is the "objective function").
Subject to:	$x_1 + x_2 \le L$	(Limit on total area).
	$F_1 \cdot x_1 + F_2 \cdot x_2 \le F$	(Limit on fertilizer).
	$P_1 \cdot x_1 + P_2 \cdot x_2 \le P$	(Limit on pesticide).
	$x_1 \ge 0, x_2 \ge 0$	(Cannot plant a negative area).

In matrix form this becomes:

Maximize $\begin{bmatrix} S_1 & S_2 \end{bmatrix} \begin{bmatrix} x_1 \\ x_2 \end{bmatrix}$

Subject to $\begin{bmatrix} 1 & 1 \\ F_1 & F_2 \\ P_1 & P_2 \end{bmatrix} \begin{bmatrix} x_1 \\ x_2 \end{bmatrix} \le \begin{bmatrix} L \\ F \\ P \end{bmatrix}, \begin{bmatrix} x_1 \\ x_2 \end{bmatrix} \ge \begin{bmatrix} 0 \\ 0 \end{bmatrix}.$

Augmented Form (Slack form)

Linear programming problems can be converted into an *augmented form* in order to apply the common form of the simplex algorithm. This form introduces non-negative *slack variables* to replace inequalities with equalities in the constraints. The problems can then be written in the following block matrix form:

Maximize z;

$$\begin{bmatrix} 1 & -\mathrm{c}^T & 0 \\ 0 & \mathrm{A} & \mathrm{I} \end{bmatrix} \begin{bmatrix} \mathrm{z} \\ \mathrm{x} \\ \mathrm{s} \end{bmatrix} = \begin{bmatrix} 0 \\ \mathrm{b} \end{bmatrix}$$

$$\mathrm{x} \ge 0, \mathrm{s} \ge 0$$

where *s* are the newly introduced slack variables, x are the decision variables, and *z* is the variable to be maximized.

Example:

The example above is converted into the following augmented form:

Maximize: $S_1 \cdot x_1 + S_2 \cdot x_2$		(Objective function)
Subject to:	$x_1 + x_2 + x_3 = L$	(Augmented constraint)
	$F_1 \cdot x_1 + F_2 \cdot x_2 + x_4 = F$	(Augmented constraint)
	$P_1 \cdot x_1 + P_2 \cdot x_2 + x_5 = P$	(Augmented constraint)
	$x_1, x_2, x_3, x_4, x_5 \ge 0.$	

where x_3, x_4, x_5 are (non-negative) slack variables, representing in this example the unused area, the amount of unused fertilizer, and the amount of unused pesticide.

In matrix form this becomes:

Maximize z;

$$\begin{bmatrix} 1 & -S_1 & -S_2 & 0 & 0 & 0 \\ 0 & 1 & 1 & 1 & 0 & 0 \\ 0 & F_1 & F_2 & 0 & 1 & 0 \\ 0 & P_1 & P_2 & 0 & 0 & 1 \end{bmatrix} \begin{bmatrix} z \\ x_1 \\ x_2 \\ x_3 \\ x_4 \\ x_5 \end{bmatrix} = \begin{bmatrix} 0 \\ L \\ F \\ P \end{bmatrix}, \begin{bmatrix} x_1 \\ x_2 \\ x_3 \\ x_4 \\ x_5 \end{bmatrix} \geq 0.$$

Duality

Every linear programming problem, referred to as a *primal* problem, can be converted into a dual problem, which provides an upper bound to the optimal value of the primal problem. In matrix form, we can express the *primal* problem as:

- Maximize $c^T x$ subject to $Ax \leq b$, $x \geq 0$;
- with the corresponding symmetric dual problem,
- Minimize $b^T y$ subject to $A^T y \geq c$, $y \geq 0$.

An alternative primal formulation is:

- Maximize $c^T x$ subject to $Ax \leq b$;
- with the corresponding asymmetric dual problem,
- Minimize $b^T y$ subject to $A^T y = c$, $y \geq 0$.

There are two ideas fundamental to duality theory. One is the fact that (for the symmetric dual) the dual of a dual linear program is the original primal linear program. Additionally, every feasible solution for a linear program gives a bound on the optimal value of the objective function of its dual. The weak duality theorem states that the objective function value of the dual at any feasible solution is always greater than or equal to the objective function value of the primal at any feasible solution. The strong duality theorem states that if the primal has an optimal solution, x^*, then the dual also has an optimal solution, y^*, and $c^T x^* = b^T y^*$.

A linear program can also be unbounded or infeasible. Duality theory tells us that if the primal is unbounded then the dual is infeasible by the weak duality theorem. Likewise, if the dual is unbounded, then the primal must be infeasible. However, it is possible for both the dual and the primal to be infeasible.

Variations

Covering/Packing Dualities

A covering LP is a linear program of the form:

Minimize: $b^T y$,

subject to: $A^T y \geq c$, $y \geq 0$,

such that the matrix A and the vectors b and c are non-negative.

The dual of a covering LP is a packing LP, a linear program of the form:

Maximize: $c^T x$,

subject to: $Ax \leq b$, $x \geq 0$,

such that the matrix A and the vectors b and c are non-negative.

Examples:

Covering and packing LPs commonly arise as a linear programming relaxation of a combinatorial problem and are important in the study of approximation algorithms. For example, the LP relaxations of the set packing problem, the independent set problem, and the matching problem are packing LPs. The LP relaxations of the set cover problem, the vertex cover problem, and the dominating set problem are also covering LPs.

Finding a fractional coloring of a graph is another example of a covering LP. In this case, there is one constraint for each vertex of the graph and one variable for each independent set of the graph.

Complementary Slackness

It is possible to obtain an optimal solution to the dual when only an optimal solution to the primal is known using the complementary slackness theorem. The theorem states:

Suppose that $x = (x_1, x_2, \ldots, x_n)$ is primal feasible and that $y = (y_1, y_2, \ldots, y_m)$ is dual feasible. Let $(w_1, w_2, \ldots, w_m)$ denote the corresponding primal slack variables, and let $(z_1, z_2, \ldots, z_n)$ denote the corresponding dual slack variables. Then **x** and **y** are optimal for their respective problems if and only if:

- $x_j z_j = 0$, for $j = 1, 2, \ldots, n$, and
- $w_i y_i = 0$, for $i = 1, 2, \ldots, m$.

So if the *i*-th slack variable of the primal is not zero, then the *i*-th variable of the dual is equal to zero. Likewise, if the *j*-th slack variable of the dual is not zero, then the *j*-th variable of the primal is equal to zero.

This necessary condition for optimality conveys a fairly simple economic principle. In standard form (when maximizing), if there is slack in a constrained primal resource (i.e., there are "leftovers"), then additional quantities of that resource must have no value. Likewise, if there is slack in the dual (shadow) price non-negativity constraint requirement, i.e., the price is not zero, then there must be scarce supplies (no "leftovers").

Theory

Existence of Optimal Solutions

Geometrically, the linear constraints define the feasible region, which is a convex polyhedron. A linear function is a convex function, which implies that every local minimum is a global minimum; similarly, a linear function is a concave function, which implies that every local maximum is a global maximum.

An optimal solution need not exist, for two reasons. First, if the constraints are inconsistent, then no feasible solution exists: For instance, the constraints $x \geq 2$ and $x \leq 1$ cannot be satisfied jointly; in this case, we say that the LP is *infeasible*. Second, when the polytope is unbounded in the direction of the gradient of the objective function (where the gradient of the objective function is the vector of the coefficients of the objective function), then no optimal value is attained because it is always possible to do better than any finite value of the objective function.

Optimal Vertices and Rays of Polyhedra

Otherwise, if a feasible solution exists and if the constraint set is bounded, then the optimum value is always attained on the boundary of the constraint set, by the *maximum principle* for *convex functions* (alternatively, by the *minimum* principle for *concave functions*) since linear functions are both convex and concave. However, some problems have distinct optimal solutions; for example, the problem of finding a feasible solution to a system of linear inequalities is a linear programming problem in which the objective function is the zero function (that is, the constant function taking the value zero everywhere). For this feasibility problem with the zero-function for its objective-function, if there are two distinct solutions, then every convex combination of the solutions is a solution.

Algorithms

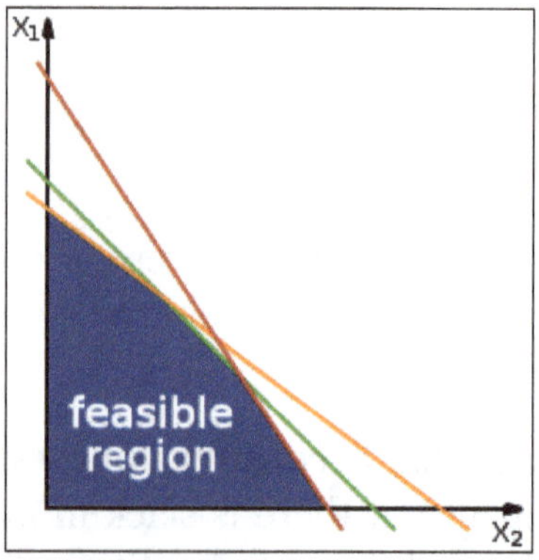

In a linear programming problem, a series of linear constraints produces a convex feasible region of possible values for those variables. In the two-variable case this region is in the shape of a convex simple polygon.

Basis Exchange Algorithms

Simplex Algorithm of Dantzig

The simplex algorithm, developed by George Dantzig in 1947, solves LP problems by constructing a feasible solution at a vertex of the polytope and then walking along a path on the edges of the polytope to vertices with non-decreasing values of the objective function until an optimum is reached for sure. In many practical problems, "stalling" occurs: many pivots are made with no increase in the objective function. In rare practical problems, the usual versions of the simplex algorithm may actually "cycle". To avoid cycles, researchers developed new pivoting rules.

In practice, the simplex algorithm is quite efficient and can be guaranteed to find the global optimum if certain precautions against *cycling* are taken. The simplex algorithm has been proved to solve "random" problems efficiently, i.e. in a cubic number of steps, which is similar to its behavior on practical problems.

However, the simplex algorithm has poor worst-case behavior: Klee and Minty constructed a family of linear programming problems for which the simplex method takes a number of steps exponential in the problem size. In fact, for some time it was not known whether the linear programming problem was solvable in polynomial time, i.e. of complexity class P.

Criss-cross Algorithm

Like the simplex algorithm of Dantzig, the criss-cross algorithm is a basis-exchange algorithm that pivots between bases. However, the criss-cross algorithm need not maintain feasibility, but can pivot rather from a feasible basis to an infeasible basis. The criss-cross algorithm does not have polynomial time-complexity for linear programming. Both algorithms visit all 2^D corners of a (perturbed) cube in dimension D, the Klee–Minty cube, in the worst case.

Interior Point

In contrast to the simplex algorithm, which finds an optimal solution by traversing the edges between vertices on a polyhedral set, interior-point methods move through the interior of the feasible region.

Ellipsoid Algorithm, Following Khachiyan

This is the first worst-case polynomial-time algorithm ever found for linear programming. To solve a problem which has n variables and can be encoded in L input bits, this algorithm uses $O(n^4L)$ pseudo-arithmetic operations on numbers with $O(L)$ digits. Leonid Khachiyan solved this long-standing complexity issue in 1979 with the introduction of the ellipsoid method. The convergence analysis has (real-number) predecessors, notably the iterative methods developed by Naum Z. Shor and the approximation algorithms by Arkadi Nemirovski and D. Yudin.

Projective Algorithm of Karmarkar

Khachiyan's algorithm was of landmark importance for establishing the polynomial-time solvability of linear programs. The algorithm was not a computational breakthrough, as the simplex method is more efficient for all but specially constructed families of linear programs.

However, Khachiyan's algorithm inspired new lines of research in linear programming. In 1984, N. Karmarkar proposed a projective method for linear programming. Karmarkar's algorithm improved on Khachiyan's worst-case polynomial bound (giving $O(n^{3.5}L)$). Karmarkar claimed that his algorithm was much faster in practical LP than the simplex method, a claim that created great interest in interior-point methods. Since Karmarkar's discovery, many interior-point methods have been proposed and analyzed.

Affine Scaling

Affine scaling is one of the oldest interior point methods to be developed. It was developed in the Soviet Union in the mid-1960s, but didn't receive much attention until the discovery of Karmarkar's algorithm, after which affine scaling was reinvented multiple times and presented as a simplified version of Karmarkar's. Affine scaling amounts to doing gradient descent steps within the feasible region, while rescaling the problem to make sure the steps move toward the optimum faster.

Vaidya's Algorithm

In 1989, Vaidya developed an algorithm that runs in $O(n^{2.5})$ time. Formally speaking, the algorithm takes $O((n+d)^{1.5}nL)$ arithmetic operations in the worst case, where d is the number of constraints, n is the number of variables, and L is the number of bits.

Path-following Algorithms

For both theoretical and practical purposes, barrier function or path-following methods have been the most popular interior point methods since the 1990s. In 2015, Lee and Sidford showed that, it can be solved in $O((nnz(A)+n^2)\sqrt{n})$ time, and it remains taking $O(n^{2.5})$ in the worst case. In 2018, Cohen, Lee and Song improved the running time to $O(n^{\omega})$ time, ω is the exponent of matrix multiplication.

Comparison of Interior-point Methods and Simplex Algorithms

The current opinion is that the efficiencies of good implementations of simplex-based methods and interior point methods are similar for routine applications of linear programming. However, for specific types of LP problems, it may be that one type of solver is better than another (sometimes much better), and that the structure of the solutions generated by interior point methods versus simplex-based methods are significantly different with the support set of active variables being typically smaller for the later one.

Approximate Algorithms for Covering/Packing LPs

Covering and packing LPs can be solved approximately in nearly-linear time. That is, if matrix A is of dimension $n \times m$ and has N non-zero entries, then there exist algorithms that run in time $O(N \cdot (\log N)^{O(1)}/\varepsilon^{O(1)})$ and produce $O(1 \pm \varepsilon)$ approximate solutions to given covering and packing LPs. The best known sequential algorithm of this kind runs in time $O(N + (\log N) \cdot (n+m)/\varepsilon^2)$, and the best known parallel algorithm of this kind runs in $O((\log N)^2/\varepsilon^3)$ iterations, each requiring only a matrix-vector multiplication which is highly parallelizable.

Open Problems and Recent Work

There are several open problems in the theory of linear programming, the solution of which would represent fundamental breakthroughs in mathematics and potentially major advances in our ability to solve large-scale linear programs.

- Does LP admit a strongly polynomial-time algorithm?
- Does LP admit a strongly polynomial-time algorithm to find a strictly complementary solution?
- Does LP admit a polynomial-time algorithm in the real number (unit cost) model of computation?

This closely related set of problems has been cited by Stephen Smale as among the 18 greatest unsolved problems of the 21st century. In Smale's words, the third version of the problem "is the main unsolved problem of linear programming theory." While algorithms exist to solve linear programming in weakly polynomial time, such as the ellipsoid methods and interior-point techniques, no algorithms have yet been found that allow strongly polynomial-time performance in the number of constraints and the number of variables. The development of such algorithms would be of great theoretical interest, and perhaps allow practical gains in solving large LPs as well.

Although the Hirsch conjecture was recently disproved for higher dimensions, it still leaves the following questions open.

- Are there pivot rules which lead to polynomial-time simplex variants?
- Do all polytopal graphs have polynomially bounded diameter?

These questions relate to the performance analysis and development of simplex-like methods. The immense efficiency of the simplex algorithm in practice despite its exponential-time theoretical performance hints that there may be variations of simplex that run in polynomial or even strongly polynomial time. It would be of great practical and theoretical significance to know whether any such variants exist, particularly as an approach to deciding if LP can be solved in strongly polynomial time.

The simplex algorithm and its variants fall in the family of edge-following algorithms, so named because they solve linear programming problems by moving from vertex to vertex along edges of a polytope. This means that their theoretical performance is limited by the maximum number of edges between any two vertices on the LP polytope. As a result, we are interested in knowing the maximum graph-theoretical diameter of polytopal graphs. It has been proved that all polytopes have subexponential diameter. The recent disproof of the Hirsch conjecture is the first step to prove whether any polytope has superpolynomial diameter. If any such polytopes exist, then no edge-following variant can run in polynomial time. Questions about polytope diameter are of independent mathematical interest.

Simplex pivot methods preserve primal (or dual) feasibility. On the other hand, criss-cross pivot methods do not preserve (primal or dual) feasibility—they may visit primal feasible, dual feasible or primal-and-dual infeasible bases in any order. Pivot methods of this type have been studied since the 1970s. Essentially, these methods attempt to find the shortest pivot path on the arrangement polytope under the linear programming problem. In contrast to polytopal graphs, graphs of arrangement polytopes are known to have small diameter, allowing the possibility of strongly polynomial-time criss-cross pivot algorithm without resolving questions about the diameter of general polytopes.

Integer Unknowns

If all of the unknown variables are required to be integers, then the problem is called an integer programming (IP) or integer linear programming (ILP) problem. In contrast to linear programming, which can be solved efficiently in the worst case, integer programming problems are in many practical situations (those with bounded variables) NP-hard. 0–1 integer programming or binary integer programming (BIP) is the special case of integer programming where variables are required to be 0 or 1 (rather than arbitrary integers). This problem is also classified as NP-hard, and in fact the decision version was one of Karp's 21 NP-complete problems.

If only some of the unknown variables are required to be integers, then the problem is called a mixed integer programming (MIP) problem. These are generally also NP-hard because they are even more general than ILP programs.

There are however some important subclasses of IP and MIP problems that are efficiently solvable, most notably problems where the constraint matrix is totally unimodular and the right-hand sides of the constraints are integers or – more general – where the system has the total dual integrality (TDI) property.

Advanced algorithms for solving integer linear programs include:

- Cutting-plane method.
- Branch and bound.
- Branch and cut.
- Branch and price.
- If the problem has some extra structure, it may be possible to apply delayed column generation.

Such integer-programming algorithms are discussed by Padberg and in Beasley.

Integral Linear Arograms

A linear program in real variables is said to be integral if it has at least one optimal solution which is integral. Likewise, a polyhedron $P = \{x \mid Ax \geq 0\}$ is said to be integral if for all bounded feasible objective functions c, the linear program $\{\max cx \mid x \in P\}$ has an optimum x^* with integer coordinates. As observed by Edmonds and Giles in 1977, one can equivalently say that the polyhedron P is integral if for every bounded feasible integral objective function c, the optimal value of the linear program $\{\max cx \mid x \in P\}$ is an integer.

Integral linear programs are of central importance in the polyhedral aspect of combinatorial optimization since they provide an alternate characterization of a problem. Specifically, for any problem, the convex hull of the solutions is an integral polyhedron; if this polyhedron has a nice/compact description, then we can efficiently find the optimal feasible solution under any linear objective. Conversely, if we can prove that a linear programming relaxation is integral, then it is the desired description of the convex hull of feasible (integral) solutions.

Terminology is not consistent throughout the literature, so one should be careful to distinguish the following two concepts:

- In an *integer linear program,* variables are forcibly constrained to be integers, and this problem is NP-hard in general.
- In an *integral linear program,* variables are not constrained to be integers but rather one has proven somehow that the continuous problem always has an integral optimal value (assuming *c* is integral), and this optimal value may be found efficiently since all polynomial-size linear programs can be solved in polynomial time.

One common way of proving that a polyhedron is integral is to show that it is totally unimodular. There are other general methods including the integer decomposition property and total dual integrality. Other specific well-known integral LPs include the matching polytope, lattice polyhedra, submodular flow polyhedra, and the intersection of 2 generalized polymatroids/g-polymatroids.

Solvers and Scripting (Programming) Languages

Permissive Licenses

Pyomo

An open-source modeling language for large-scale linear, mixed integer and nonlinear optimization.

Copyleft (Reciprocal) Licenses

Cassowary Constraint Solver

An incremental constraint solving toolkit that efficiently solves systems of linear equalities and inequalities.

CLP

An LP solver from COIN-OR.

Glpk

GNU Linear Programming Kit, an LP/MILP solver with a native C API and numerous (15) third-party wrappers for other languages. Specialist support for flow networks. Bundles the AMPL-like GNU MathProg modelling language and translator.

Qoca

A library for incrementally solving systems of linear equations with various goal functions.

R-project

A programming language and software environment for statistical computing and graphics.

MINTO (Mixed Integer Optimizer, an integer programming solver which uses branch and bound algorithm) has publicly available source code but is not open source.

Proprietary Licenses

AIMMS

A modeling language that allows to model linear, mixed integer, and nonlinear optimization models. It also offers a tool for constraint programming. Algorithm, in the forms of heuristics or exact methods, such as Branch-and-Cut or Column Generation, can also be implemented. The tool calls an appropriate solver such as CPLEX, Gurobi or similar, to solve the optimization problem at hand. Academic licenses are free of charge.

AMPL

A popular modeling language for large-scale linear, mixed integer and nonlinear optimisation with a free student limited version available (500 variables and 500 constraints).

APMonitor

API to MATLAB and Python. Solve example Linear Programming (LP) problems through MATLAB, Python, or a web-interface.

CPLEX

Popular solver with an API for several programming languages, and also has a modelling language and works with AIMMS, AMPL, GAMS, MPL, OpenOpt, OPL Development Studio, and TOMLAB.

Excel Solver Function

A nonlinear solver adjusted to spreadsheets in which function evaluations are based on the recalculating cells. Basic version available as a standard add-on for Excel.

Gurobi

Solver with parallel algorithms for large-scale linear programs, quadratic programs and mixed-integer programs.

IMSL Numerical Libraries

Collections of math and statistical algorithms available in C/C++, Fortran, Java and C#/.NET. Optimization routines in the IMSL Libraries include unconstrained, linearly and nonlinearly constrained minimizations, and linear programming algorithms.

LINDO

Solver with an API for large scale optimization of linear, integer, quadratic, conic and general nonlinear programs with stochastic programming extensions. It offers a global optimization procedure for finding guaranteed globally optimal solution to general nonlinear programs with continuous and discrete variables. It also has a statistical sampling API to integrate Monte-Carlo simulations into an optimization framework. It has an algebraic modeling language (LINGO) and allows modeling within a spreadsheet.

Maple

A general-purpose programming-language for symbolic and numerical computing.

MATLAB

A general-purpose and matrix-oriented programming-language for numerical computing. Linear programming in MATLAB requires the Optimization Toolbox in addition to the base MATLAB product; available routines include INTLINPROG and LINPROG.

Mathcad

A WYSIWYG math editor. It has functions for solving both linear and nonlinear optimization problems.

Mathematica

A general-purpose programming-language for mathematics, including symbolic and numerical capabilities.

MOSEK

A solver for large scale optimization with API for several languages (C++,java,.net, Matlab and python).

NAG Numerical Library

A collection of mathematical and statistical routines developed by the Numerical Algorithms Group for multiple programming languages (C, C++, Fortran, Visual Basic, Java and C#) and packages (MATLAB, Excel, R, LabVIEW). The Optimization chapter of the NAG Library includes routines for linear programming problems with both sparse and non-sparse linear constraint matrices, together with routines for the optimization of quadratic, nonlinear, sums of squares of linear or nonlinear functions with nonlinear, bounded or no constraints. The NAG Library has routines for both local and global optimization, and for continuous or integer problems.

NMath Stats

A general-purpose .NET statistical library containing a simplex solver.

OptimJ

A Java-based modeling language for optimization with a free version available.

SAS/OR

A suite of solvers for Linear, Integer, Nonlinear, Derivative-Free, Network, Combinatorial and Constraint Optimization; the Algebraic modeling language OPTMODEL; and a variety of vertical solutions aimed at specific problems/markets, all of which are fully integrated with the SAS System.

SCIP

A general-purpose constraint integer programming solver with an emphasis on MIP. Compatible with Zimpl modelling language.

XPRESS

Solver for large-scale linear programs, quadratic programs, general nonlinear and mixed-integer programs. Has API for several programming languages, also has a modelling language Mosel and works with AMPL, GAMS.

VisSim

A visual block diagram language for simulation of dynamical systems.

NONLINEAR PROGRAMMING

In mathematics, nonlinear programming (NLP) is the process of solving an optimization problem where some of the constraints or the objective function are nonlinear. An optimization problem is one of calculation of the extrema (maxima, minima or stationary points) of an objective function over a set of unknown real variables and conditional to the satisfaction of a system of equalities and inequalities, collectively termed constraints. It is the sub-field of mathematical optimization that deals with problems that are not linear.

Applicability

A typical non-convex problem is that of optimizing transportation costs by selection from a set of transportation methods, one or more of which exhibit economies of scale, with various connectivities and capacity constraints. An example would be petroleum product transport given a selection or combination of pipeline, rail tanker, road tanker, river barge, or coastal tankship. Owing to economic batch size the cost functions may have discontinuities in addition to smooth changes.

In experimental science, some simple data analysis (such as fitting a spectrum with a sum of peaks of known location and shape but unknown magnitude) can be done with linear methods, but in general these problems, also, are nonlinear. Typically, one has a theoretical model of the system under study with variable parameters in it and a model the experiment or experiments, which may also have unknown parameters. One tries to find a best fit numerically. In this case one often wants a measure of the precision of the result, as well as the best fit itself.

Let n, m, and p be positive integers. Let X be a subset of R^n, let f, g_i, and h_j be real-valued functions on X for each i in $\{1, ..., m\}$ and each j in $\{1, ..., p\}$, with at least one of f, g_i, and h_j being nonlinear.

A nonlinear minimization problem is an optimization problem of the form:

$$\begin{aligned}
&\text{Minimize}: f(x) \\
&\text{Subject to}: g_i(x) \leq 0 \text{ for each } i \in \{1,...,m\} \\
&\qquad h_j(x) = 0 \text{ for each } j \in \{1,..,p\} \\
&\qquad x \in X.
\end{aligned}$$

A nonlinear maximization problem is defined in a similar way.

Possible Types of Constraint Set

There are several possibilities for the nature of the constraint set, also known as the feasible set or feasible region.

An *infeasible* problem is one for which no set of values for the choice variables satisfies all the constraints. That is, the constraints are mutually contradictory, and no solution exists; the feasible set is the empty set.

A *feasible* problem is one for which there exists at least one set of values for the choice variables satisfying all the constraints.

An *unbounded* problem is a feasible problem for which the objective function can be made to be better than any given finite value. Thus there is no optimal solution, because there is always a feasible solution that gives a better objective function value than does any given proposed solution.

Methods for Solving the Problem

If the objective function f is linear and the constrained space is a polytope, the problem is a linear programming problem, which may be solved using well-known linear programming techniques such as the simplex method.

If the objective function is concave (maximization problem), or convex (minimization problem) and the constraint set is convex, then the program is called convex and general methods from convex optimization can be used in most cases.

If the objective function is quadratic and the constraints are linear, quadratic programming techniques are used.

If the objective function is a ratio of a concave and a convex function (in the maximization case) and the constraints are convex, then the problem can be transformed to a convex optimization problem using fractional programming techniques.

Several methods are available for solving nonconvex problems. One approach is to use special formulations of linear programming problems. Another method involves the use of branch and bound techniques, where the program is divided into subclasses to be solved with convex (minimization problem) or linear approximations that form a lower bound on the overall cost within the subdivision. With subsequent divisions, at some point an actual solution will be obtained whose cost is equal to the best lower bound obtained for any of the approximate solutions. This solution is optimal, although possibly not unique. The algorithm may also be stopped early, with the assurance that the best possible solution is within a tolerance from the best point found; such points are called ε-optimal. Terminating to ε-optimal points is typically necessary to ensure finite termination. This is especially useful for large, difficult problems and problems with uncertain costs or values where the uncertainty can be estimated with an appropriate reliability estimation.

Under differentiability and constraint qualifications, the Karush–Kuhn–Tucker (KKT) conditions provide necessary conditions for a solution to be optimal. Under convexity, these conditions are also sufficient. If some of the functions are non-differentiable, subdifferential versions of Karush–Kuhn–Tucker (KKT) conditions are available.

Examples

2-dimensional Example

A simple problem can be defined by the constraints:

$$x_1 \geq 0$$
$$x_2 \geq 0$$
$$x_1^2 + x_2^2 \geq 1$$
$$x_1^2 + x_2^2 \leq 2$$

with an objective function to be maximized:

$$f(x) = x_1 + x_2$$

where $x = (x_1, x_2)$.

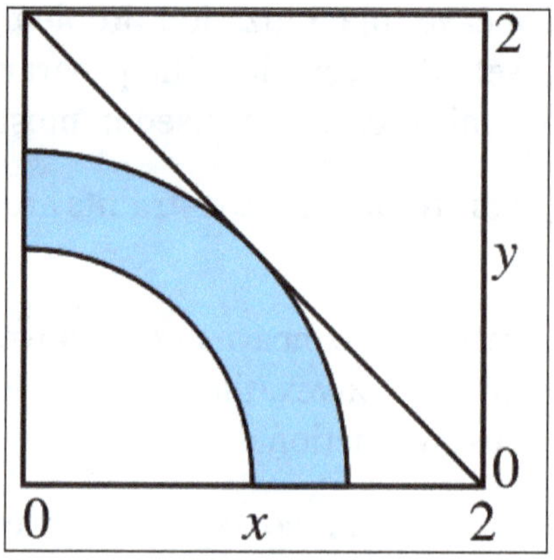

The blue region is the feasible region. The tangency of the line with the feasible region represents the solution. The line is the best achievable contour line (locus with a given value of the objective function).

3-dimensional Example

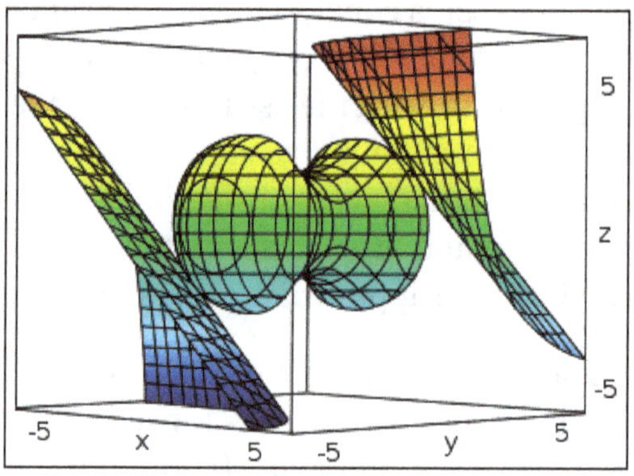

The tangency of the top surface with the constrained space in the center represents the solution.

Another simple problem can be defined by the constraints:

$$x_1^2 - x_2^2 + x_3^2 \leq 2$$
$$x_1^2 + x_2^2 + x_3^2 \leq 10$$

with an objective function to be maximized:

$$f(\mathrm{x}) = x_1 x_2 + x_2 x_3$$

where $\mathrm{x} = (x_1, x_2, x_3)$.

DYNAMIC PROGRAMMING

Dynamic programming is both a mathematical optimization method and a computer programming method. The method was developed by Richard Bellman in the 1950s and has found applications in numerous fields, from aerospace engineering to economics. In both contexts it refers to simplifying a complicated problem by breaking it down into simpler sub-problems in a recursive manner. While some decision problems cannot be taken apart this way, decisions that span several points in time do often break apart recursively. Likewise, in computer science, if a problem can be solved optimally by breaking it into sub-problems and then recursively finding the optimal solutions to the sub-problems, then it is said to have optimal substructure.

If sub-problems can be nested recursively inside larger problems, so that dynamic programming methods are applicable, then there is a relation between the value of the larger problem and the values of the sub-problems. In the optimization literature this relationship is called the Bellman equation

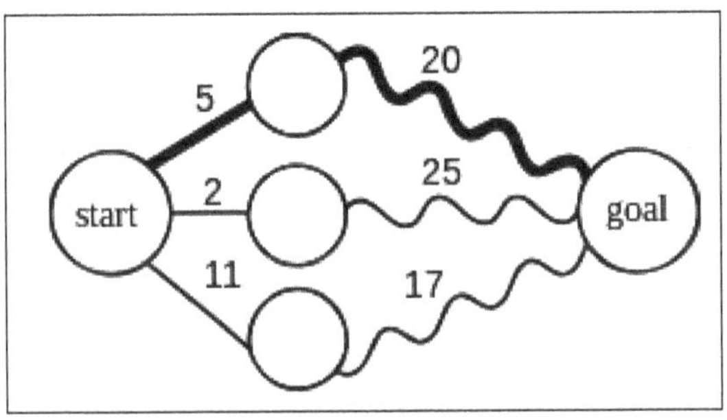

Finding the shortest path in a graph using optimal substructure; a straight line indicates a single edge; a wavy line indicates a shortest path between the two vertices it connects (among other paths, not shown, sharing the same two vertices); the bold line is the overall shortest path from start to goal.

Mathematical Optimization

In terms of mathematical optimization, dynamic programming usually refers to simplifying a decision by breaking it down into a sequence of decision steps over time. This is done by defining a sequence of value functions $V_1, V_2, \ldots, V_n$ taking y as an argument representing the state of the system at times i from 1 to n. The definition of $V_n(y)$ is the value obtained in state y at the last time n. The values V_i at earlier times i = n −1, n − 2, ..., 2, 1 can be found by working backwards, using a recursive relationship called the Bellman equation. For i = 2, ..., n, V_{i-1} at any state y is calculated from V_i by maximizing a simple function (usually the sum) of the gain from a decision at time $i - 1$ and the function V_i at the new state of the system if this decision is made. Since V_i has already been calculated for the needed states, the above operation yields V_{i-1} for those states. Finally, V_1 at the initial state of the system is the value of the optimal solution. The optimal values of the decision variables can be recovered, one by one, by tracking back the calculations already performed.

Computer Programming

There are two key attributes that a problem must have in order for dynamic programming to be applicable: optimal substructure and overlapping sub-problems. If a problem can be solved by combining optimal solutions to *non-overlapping* sub-problems, the strategy is called "divide and conquer" instead. This is why merge sort and quick sort are not classified as dynamic programming problems.

Optimal substructure means that the solution to a given optimization problem can be obtained by the combination of optimal solutions to its sub-problems. Such optimal substructures are usually described by means of recursion. For example, given a graph *G=(V,E)*, the shortest path *p* from a vertex *u* to a vertex *v* exhibits optimal substructure: take any intermediate vertex *w* on this shortest path *p*. If *p* is truly the shortest path, then it can be split into sub-paths p_1 from *u* to *w* and p_2 from *w* to *v* such that these, in turn, are indeed the shortest paths between the corresponding vertices (by the simple cut-and-paste argument described in *Introduction to Algorithms*). Hence, one can easily formulate the solution for finding shortest paths in a recursive manner, which is what the Bellman–Ford algorithm or the Floyd–Warshall algorithm does.

Overlapping sub-problems means that the space of sub-problems must be small, that is, any recursive algorithm solving the problem should solve the same sub-problems over and over, rather than generating new sub-problems. For example, consider the recursive formulation for generating the Fibonacci series: $F_i = F_{i-1} + F_{i-2}$, with base case $F_1 = F_2 = 1$. Then $F_{43} = F_{42} + F_{41}$, and $F_{42} = F_{41} + F_{40}$. Now F_{41} is being solved in the recursive sub-trees of both F_{43} as well as F_{42}. Even though the total number of sub-problems is actually small (only 43 of them), we end up solving the same problems over and over if we adopt a naive recursive solution such as this. Dynamic programming takes account of this fact and solves each sub-problem only once.

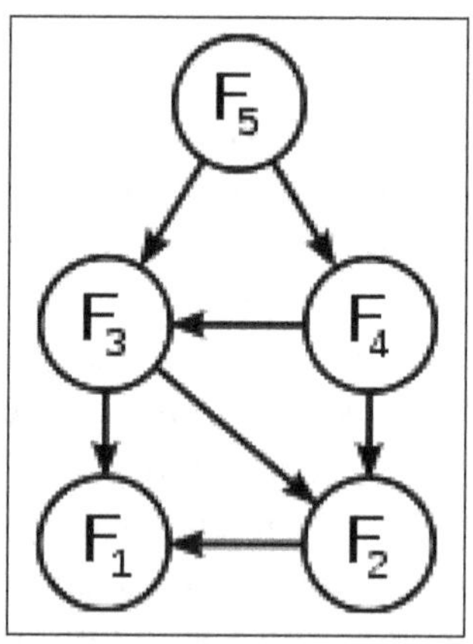

The subproblem graph for the Fibonacci sequence. The fact that it is not a tree indicates overlapping subproblems.

This can be achieved in either of two ways:

- Top-down approach: This is the direct fall-out of the recursive formulation of any problem. If the solution to any problem can be formulated recursively using the solution to its sub-problems, and if its sub-problems are overlapping, then one can easily memoize or store the solutions to the sub-problems in a table.

Whenever we attempt to solve a new sub-problem, we first check the table to see if it is already solved. If a solution has been recorded, we can use it directly, otherwise we solve the sub-problem and add its solution to the table.

- Bottom-up approach: Once we formulate the solution to a problem recursively as in terms of its sub-problems, we can try reformulating the problem in a bottom-up fashion: try solving the sub-problems first and use their solutions to build-on and arrive at solutions to bigger sub-problems. This is also usually done in a tabular form by iteratively generating solutions to bigger and bigger sub-problems by using the solutions to small sub-problems. For example, if we already know the values of F_{41} and F_{40}, we can directly calculate the value of F_{42}.

Some programming languages can automatically memoize the result of a function call with a particular set of arguments, in order to speed up call-by-name evaluation (this mechanism is referred to as *call-by-need*). Some languages make it possible portably (e.g. Scheme, Common Lisp, Perl or D). Some languages have automatic memoization built in, such as tabled Prolog and J, which supports memoization with the *M.* adverb. In any case, this is only possible for a referentially transparent function. Memoization is also encountered as an easily accessible design pattern within term-rewrite based languages such as Wolfram Language.

Computer Algorithms

Dijkstra's Algorithm for the Shortest Path Problem

From a dynamic programming point of view, Dijkstra's algorithm for the shortest path problem is a successive approximation scheme that solves the dynamic programming functional equation for the shortest path problem by the reaching method.

In fact, Dijkstra's explanation of the logic behind the algorithm, namely is a paraphrasing of Bellman's famous Principle of Optimality in the context of the shortest path problem.

Find the path of minimum total length between two given nodes p and Q.

We use the fact that, if R is a node on the minimal path from P to Q, knowledge of the latter implies the knowledge of the minimal path from P to R.

Fibonacci Sequence

Here is a naïve implementation of a function finding the nth member of the Fibonacci sequence, based directly on the mathematical definition:

```
function fib(n)

    if n <= 1 return n
```

```
        return fib(n - 1) + fib(n - 2)
```

Notice that if we call, say, fib(5), we produce a call tree that calls the function on the same value many different times:

1. fib(5)
2. fib(4) + fib(3)
3. (fib(3) + fib(2)) + (fib(2) + fib(1))
4. ((fib(2) + fib(1)) + (fib(1) + fib(0))) + ((fib(1) + fib(0)) + fib(1))
5. (((fib(1) + fib(0)) + fib(1)) + (fib(1) + fib(0))) + ((fib(1) + fib(0)) + fib(1))

In particular, fib(2) was calculated three times from scratch. In larger examples, many more values of fib, or *subproblems*, are recalculated, leading to an exponential time algorithm.

Now, suppose we have a simple map object, *m*, which maps each value of fib that has already been calculated to its result, and we modify our function to use it and update it. The resulting function requires only O(n) time instead of exponential time (but requires O(n) space):

```
var m := map(0 → 0, 1 → 1)

function fib(n)

    if key n is not in map m

        m[n] := fib(n - 1) + fib(n - 2)

    return m[n]
```

This technique of saving values that have already been calculated is called *memoization*; this is the top-down approach, since we first break the problem into subproblems and then calculate and store values.

In the bottom-up approach, we calculate the smaller values of fib first, then build larger values from them. This method also uses O(n) time since it contains a loop that repeats n − 1 times, but it only takes constant (O(1)) space, in contrast to the top-down approach which requires O(n) space to store the map.

```
function fib(n)

    if n = 0

        return 0

    else
```

```
        var previousFib := 0, currentFib := 1
        repeat n - 1 times // loop is skipped if n = 1
            var newFib := previousFib + currentFib
            previousFib := currentFib
            currentFib  := newFib
    return currentFib
```

We only calculate fib(2) one time, and then use it to calculate both fib(4) and fib(3), instead of computing it every time either of them is evaluated.

The described method actually takes $\Omega(n^2)$ time for large n because addition of two integers with $\Omega(n)$ bits each takes $\Omega(n)$ time. (The nth fibonacci number has $\Omega(n)$ bits.) Also, there is a closed form for the Fibonacci sequence, known as Binet's formula, from which the n-th term can be computed in approximately $O(n(\log n)^2)$ time, which is more efficient than the described dynamic programming technique. However, the simple recurrence directly gives the matrix form that leads to an approximately $O(n \log n)$ algorithm by fast matrix exponentiation.

Balanced 0–1 Matrix

Consider the problem of assigning values, either zero or one, to the positions of an $n \times n$ matrix, with n even, so that each row and each column contains exactly $n / 2$ zeros and $n / 2$ ones. We ask how many different assignments there are for a given . For example, when $n = 4$, four possible solutions are:

$$\begin{bmatrix} 0 & 1 & 0 & 1 \\ 1 & 0 & 1 & 0 \\ 0 & 1 & 0 & 1 \\ 1 & 0 & 1 & 0 \end{bmatrix} \text{ and } \begin{bmatrix} 0 & 0 & 1 & 1 \\ 0 & 0 & 1 & 1 \\ 1 & 1 & 0 & 0 \\ 1 & 1 & 0 & 0 \end{bmatrix} \text{ and } \begin{bmatrix} 1 & 1 & 0 & 0 \\ 0 & 0 & 1 & 1 \\ 1 & 1 & 0 & 0 \\ 0 & 0 & 1 & 1 \end{bmatrix} \text{ and } \begin{bmatrix} 1 & 0 & 0 & 1 \\ 0 & 1 & 1 & 0 \\ 0 & 1 & 1 & 0 \\ 1 & 0 & 0 & 1 \end{bmatrix}.$$

There are at least three possible approaches: brute force, backtracking, and dynamic programming.

Brute force consists of checking all assignments of zeros and ones and counting those that have balanced rows and columns ($n/2$ zeros and $n/2$ ones). As there are $\binom{n}{n/2}^n$ possible assignments, this strategy is not practical except maybe up to $n = 6$.

Backtracking for this problem consists of choosing some order of the matrix elements and recursively placing ones or zeros, while checking that in every row and column the number of elements that have not been assigned plus the number of ones or zeros are both at least $n/2$. While more sophisticated than brute force, this approach will visit every solution once, making it impractical for n larger than six, since the number of solutions is already 116,963,796,250 for $n = 8$.

Dynamic programming makes it possible to count the number of solutions without visiting them all. Imagine backtracking values for the first row – what information would we require about the remaining rows, in order to be able to accurately count the solutions obtained for each first row value? We consider $k \times n$ boards, where $1 \le k \le n$, whose k rows contain $n/2$ zeros and $n/2$ ones. The function f to which memoization is applied maps vectors of n pairs of integers to the number of admissible boards (solutions). There is one pair for each column, and its two components indicate respectively the number of zeros and ones that have yet to be placed in that column. We seek the value of $f((n/2,n/2),(n/2,n/2),\ldots(n/2,n/2))$ (n arguments or one vector of n elements). The process of subproblem creation involves iterating over every one of $\binom{n}{n/2}$ possible assignments for the top row of the board, and going through every column, subtracting one from the appropriate element of the pair for that column, depending on whether the assignment for the top row contained a zero or a one at that position. If any one of the results is negative, then the assignment is invalid and does not contribute to the set of solutions (recursion stops). Otherwise, we have an assignment for the top row of the $k \times n$ board and recursively compute the number of solutions to the remaining $(k-1) \times n$ board, adding the numbers of solutions for every admissible assignment of the top row and returning the sum, which is being memoized. The base case is the trivial subproblem, which occurs for a $1 \times n$ board. The number of solutions for this board is either zero or one, depending on whether the vector is a permutation of $n/2$ $(0,1)$ and $n/2$ $(1,0)$ pairs or not.

For example, in the first two boards shown above the sequences of vectors would be:

((2, 2) (2, 2) (2, 2) (2, 2))	((2, 2) (2, 2) (2, 2) (2, 2))	k = 4
0 1 0 1	0 0 1 1	
((1, 2) (2, 1) (1, 2) (2, 1))	((1, 2) (1, 2) (2, 1) (2, 1))	k = 3
1 0 1 0	0 0 1 1	
((1, 1) (1, 1) (1, 1) (1, 1))	((0, 2) (0, 2) (2, 0) (2, 0))	k = 2
0 1 0 1	1 1 0 0	
((0, 1) (1, 0) (0, 1) (1, 0))	((0, 1) (0, 1) (1, 0) (1, 0))	k = 1
1 0 1 0	1 1 0 0	

```
((0, 0) (0, 0) (0, 0) (0, 0))          ((0, 0) (0, 0), (0, 0) (0, 0))
```

The number of solutions (sequence A058527 in the OEIS) is:

$$1, 2, 90, 297200, 116963796250, 6736218287430460752, \ldots$$

Links to the MAPLE implementation of the dynamic programming approach may be found among the external links.

Checkerboard

Consider a checkerboard with $n \times n$ squares and a cost function c(i, j) which returns a cost associated with square (i,j) (*i* being the row, *j* being the column). For instance (on a 5 × 5 checkerboard),

5	6	7	4	7	8
4	7	6	1	1	4
3	3	5	7	8	2
2	–	6	7	0	–
1	–	–	*5*	–	–
	1	2	3	4	5

Thus c(1, 3) = 5.

Let us say there was a checker that could start at any square on the first rank (i.e., row) and you wanted to know the shortest path (the sum of the minimum costs at each visited rank) to get to the last rank; assuming the checker could move only diagonally left forward, diagonally right forward, or straight forward. That is, a checker on (1,3) can move to (2,2), (2,3) or (2,4).

5					
4					
3					
2		x	x	x	
1			o		
	1	2	3	4	5

This problem exhibits optimal substructure. That is, the solution to the entire problem relies on solutions to subproblems. Let us define a function q(i, j) as:

$q(i, j)$ = the minimum cost to reach square (i, j).

Starting at rank n and descending to rank 1, we compute the value of this function for all the squares at each successive rank. Picking the square that holds the minimum value at each rank gives us the shortest path between rank n and rank 1.

The function q(i, j) is equal to the minimum cost to get to any of the three squares below it (since those are the only squares that can reach it) plus c(i, j). For instance:

5					
4			A		
3		B	C	D	
2					
1					
	1	2	3	4	5

$$q(A) = \min(q(B), q(C), q(D)) + c(A)$$

Now, let us define q(i, j) in somewhat more general terms:

$$q(i,j) = \begin{cases} \infty & j < 1 \text{ or } j > n \\ c(i,j) & i = 1 \\ \min(q(i-1,j-1), q(i-1,j), q(i-1,j+1)) + c(i,j) & \text{otherwise.} \end{cases}$$

The first line of this equation deals with a board modeled as squares indexed on 1 at the lowest bound and n at the highest bound. The second line specifies what happens at the last rank; providing a base case. The third line, the recursion, is the important part. It represents the A,B,C,D terms in the example. From this definition we can derive straightforward recursive code for q(i, j). In the following pseudocode, n is the size of the board, c(i, j) is the cost function, and min() returns the minimum of a number of values:

```
function minCost(i, j)

    if j < 1 or j > n

        return infinity

    else if i = 1

        return c(i, j)

    else

         return min( minCost(i-1, j-1), minCost(i-1, j), minCost(i-1,
j+1) ) + c(i, j)
```

This function only computes the path cost, not the actual path. We discuss the actual path below. This, like the Fibonacci-numbers example, is horribly slow because it too exhibits the overlapping sub-problems attribute. That is, it recomputes the same path costs over and over. However, we can compute it much faster in a bottom-up fashion if we store path costs in a two-dimensional array q[i, j] rather than using a function. This avoids recomputation; all the values needed for array q[i, j] are computed ahead of time only once. Precomputed values for (i,j) are simply looked-up whenever needed.

We also need to know what the actual shortest path is. To do this, we use another array p[i, j]; a *predecessor array*. This array records the path to any square s. The predecessor of s is modeled as an offset relative to the index (in q[i, j]) of the precomputed path cost of s. To reconstruct the complete path, we lookup the predecessor of s, then the predecessor of that square, then the predecessor of that square, and so on recursively, until we reach the starting square. Consider the following code:

```
function computeShortestPathArrays()
    for x from 1 to n
        q[1, x] := c(1, x)
    for y from 1 to n
        q[y, 0]     := infinity
        q[y, n + 1] := infinity
    for y from 2 to n
        for x from 1 to n
            m := min(q[y-1, x-1], q[y-1, x], q[y-1, x+1])
            q[y, x] := m + c(y, x)
            if m = q[y-1, x-1]
                p[y, x] := -1
            else if m = q[y-1, x]
                p[y, x] :=  0
            else
                p[y, x] :=  1
```

Now the rest is a simple matter of finding the minimum and printing it.

```
function computeShortestPath()
    computeShortestPathArrays()
    minIndex := 1
    min := q[n, 1]
    for i from 2 to n
        if q[n, i] < min
            minIndex := i
            min := q[n, i]
    printPath(n, minIndex)
```

```
function printPath(y, x)
    print(x)
    print("<-")
    if y = 2
        print(x + p[y, x])
    else
        printPath(y-1, x + p[y, x])
```

Sequence Alignment

In genetics, sequence alignment is an important application where dynamic programming is essential. Typically, the problem consists of transforming one sequence into another using edit operations that replace, insert, or remove an element. Each operation has an associated cost, and the goal is to find the sequence of edits with the lowest total cost.

The problem can be stated naturally as a recursion, a sequence A is optimally edited into a sequence B by either:

1. Inserting the first character of B, and performing an optimal alignment of A and the tail of B.
2. Deleting the first character of A, and performing the optimal alignment of the tail of A and B.
3. Replacing the first character of A with the first character of B, and performing optimal alignments of the tails of A and B.

The partial alignments can be tabulated in a matrix, where cell (i,j) contains the cost of the optimal alignment of A[1..i] to B[1..j]. The cost in cell (i,j) can be calculated by adding the cost of the relevant operations to the cost of its neighboring cells, and selecting the optimum.

Tower of Hanoi Puzzle

A model set of the Towers of Hanoi (with 8 disks).

An animated solution of the Tower of Hanoi puzzle for T(4,3).

The Tower of Hanoi or Towers of Hanoi is a mathematical game or puzzle. It consists of three rods, and a number of disks of different sizes which can slide onto any rod. The puzzle starts with the disks in a neat stack in ascending order of size on one rod, the smallest at the top, thus making a conical shape.

The objective of the puzzle is to move the entire stack to another rod, obeying the following rules:

- Only one disk may be moved at a time.
- Each move consists of taking the upper disk from one of the rods and sliding it onto another rod, on top of the other disks that may already be present on that rod.
- No disk may be placed on top of a smaller disk.

The dynamic programming solution consists of solving the functional equation:

S(n,h,t) = S(n-1,h, not(h,t)) ; S(1,h,t) ; S(n-1,not(h,t),t)

where n denotes the number of disks to be moved, h denotes the home rod, t denotes the target rod, not(h,t) denotes the third rod (neither h nor t), ";" denotes concatenation, and:

> S(n, h, t) := solution to a problem consisting of n disks that are to be moved from rod h to rod t.

For n=1 the problem is trivial, namely S(1, h, t) = "move a disk from rod h to rod t" (there is only one disk left).

The number of moves required by this solution is $2^n - 1$. If the objective is to maximize the number of moves (without cycling) then the dynamic programming functional equation is slightly more complicated and $3^n - 1$ moves are required.

Egg Dropping Puzzle

The following is a description of the instance of this famous puzzle involving N=2 eggs and a building with H=36 floors:

> Suppose that we wish to know which stories in a 36-story building are safe to

drop eggs from, and which will cause the eggs to break on landing (using U.S. English terminology, in which the first floor is at ground level). We make a few assumptions:

- An egg that survives a fall can be used again.
- A broken egg must be discarded.
- The effect of a fall is the same for all eggs.
- If an egg breaks when dropped, then it would break if dropped from a higher window.
- If an egg survives a fall, then it would survive a shorter fall.
- It is not ruled out that the first-floor windows break eggs, nor is it ruled out that eggs can survive the 36th-floor windows.

If only one egg is available and we wish to be sure of obtaining the right result, the experiment can be carried out in only one way. Drop the egg from the first-floor window; if it survives, drop it from the second-floor window. Continue upward until it breaks. In the worst case, this method may require 36 droppings. Suppose 2 eggs are available. What is the lowest number of egg-droppings that is guaranteed to work in all cases?

To derive a dynamic programming functional equation for this puzzle, let the state of the dynamic programming model be a pair s = (n,k), where:

- n = number of test eggs available, $n = 0, 1, 2, 3, \ldots, N - 1$.
- k = number of (consecutive) floors yet to be tested, $k = 0, 1, 2, \ldots, H - 1$.

For instance, $s = (2,6)$ indicates that two test eggs are available and 6 (consecutive) floors are yet to be tested. The initial state of the process is $s = (N,H)$ where N denotes the number of test eggs available at the commencement of the experiment. The process terminates either when there are no more test eggs ($n = 0$) or when $k = 0$, whichever occurs first. If termination occurs at state $s = (0,k)$ and $k > 0$, then the test failed.

Now, let:

> $W(n,k)$ = minimum number of trials required to identify the value of the critical floor under the worst-case scenario given that the process is in state $s = (n,k)$.

Then it can be shown that:

$$W(n,k) = 1 + \min\{\max(W(n-1, x-1), W(n,k-x)): x = 1, 2, \ldots, k\}$$

with $W(n,0) = 0$ for all $n > 0$ and $W(1,k) = k$ for all k. It is easy to solve this equation iteratively by systematically increasing the values of n and k.

An interactive online facility is available for experimentation with this model as well as with other versions of this puzzle (e.g., when the objective is to minimize the expected value of the number of trials).

Faster DP Solution using a Different Parametrization

Notice that the above solution takes $O(nk^2)$ time with a DP solution. This can be improved to $O(nk\log k)$ time by binary searching on the optimal x in the above recurrence, since $W(n-1,x-1)$ is increasing in x while $W(n,k-x)$ is decreasing in x, thus a local minimum of $\max(W(n-1,x-1),W(n,k-x))$ is a global minimum. Also, by storing the optimal x for each cell in the DP table and referring to its value for the previous cell, the optimal x for each cell can be found in constant time, improving it to $O(nk)$ time. However, there is an even faster solution that involves a different parametrization of the problem:

Let k be the total number of floors such that the eggs break when dropped from the kth floor (The example above is equivalent to taking $k=37$).

Let m be the minimum floor from which the egg must be dropped to be broken.

Let $f(t,n)$ be the maximum number of values of m that are distinguishable using t tries and n eggs.

Then $f(t,0)=f(0,n)=1$ for all $t,n\geq 0$.

Let $f(t,n)$ be the floor from which the first egg is dropped in the optimal strategy.

If the first egg broke, m is from 1to a and distinguishable using at most $t-1$ tries and *n-1* eggs.

If the first egg did not break, m is from $a+1$ to k and distinguishable using $t-1$ tries and n eggs.

Therefore, $f(t,n)=f(t-1,n-1)+f(t-1,n)$.

Then the problem is equivalent to finding the minimum x such that $f(x,n)\geq k$.

To do so, we could compute $\{f(t,i): 0\leq i\leq n\}$ in order of increasing t, which would take $O(nx)$ time.

Thus, if we separately handle the case of $n=1$, the algorithm would take $O(n\sqrt{k})$ time.

But the recurrence relation can in fact be solved, giving $f(t,n)=\sum_{i=0}^{n}\binom{t}{i}$, which can be computed in $O(n)$ time using the identity $\binom{t}{i+1}=\binom{t}{i}\frac{t-i}{i+1}f$ for all $i\geq 0$.

Since $f(t,n)\leq f(t+1,n)$ for all $t\geq 0$, we can binary search on t to find x, giving an $O(n\log k)$ algorithm.

Matrix Chain Multiplication

Matrix chain multiplication is a well-known example that demonstrates utility of dynamic programming. For example, engineering applications often have to multiply a chain of matrices. It is not surprising to find matrices of large dimensions, for example 100×100. Therefore, our task is to multiply matrices $A_1, A_2, \ldots, A_n$. As we know from basic linear algebra, matrix multiplication is not commutative, but is associative; and we can multiply only two matrices at a time. So, we can multiply this chain of matrices in many different ways, for example:

$$((A_1 \times A_2) \times A_3) \times \ldots A_n$$

$$A_1 \times (((A_2 \times A_3) \times \ldots) \times A_n)$$

$$(A_1 \times A_2) \times (A_3 \times \ldots A_n)$$

and so on. There are numerous ways to multiply this chain of matrices. They will all produce the same final result, however they will take more or less time to compute, based on which particular matrices are multiplied. If matrix A has dimensions m×n and matrix B has dimensions n×q, then matrix C=A×B will have dimensions m×q, and will require m*n*q scalar multiplications (using a simplistic matrix multiplication algorithm for purposes of illustration).

For example, let us multiply matrices A, B and C. Let us assume that their dimensions are m×n, n×p, and p×s, respectively. Matrix A×B×C will be of size m×s and can be calculated in two ways shown below:

1. Ax(B×C) This order of matrix multiplication will require nps + mns scalar multiplications.
2. (A×B)×C This order of matrix multiplication will require mnp + mps scalar calculations.

Let us assume that m = 10, n = 100, p = 10 and s = 1000. So, the first way to multiply the chain will require 1,000,000 + 1,000,000 calculations. The second way will require only 10,000+100,000 calculations. Obviously, the second way is faster, and we should multiply the matrices using that arrangement of parenthesis.

Therefore, our conclusion is that the order of parenthesis matters, and that our task is to find the optimal order of parenthesis.

At this point, we have several choices, one of which is to design a dynamic programming algorithm that will split the problem into overlapping problems and calculate the optimal arrangement of parenthesis. The dynamic programming solution is presented below.

Let's call m[i,j] the minimum number of scalar multiplications needed to multiply a chain of matrices from matrix i to matrix j (i.e. $A_i \times .... \times A_j$, i.e. i<=j). We split the chain at some matrix k, such that i <= k < j, and try to find out which combination produces minimum m[i,j].

The formula is:

```
        if i = j, m[i,j]= 0

      if i < j, m[i,j]= min over all possible values of k (m[i,k]+m[k+1,j]
+ p_{i-1} * p_k * p_j)
```

where k ranges from i to $j - 1$.

- p_{i-1} is the row dimension of matrix i,
- p_k is the column dimension of matrix k,
- p_j is the column dimension of matrix j.

This formula can be coded as shown below, where input parameter "chain" is the chain of matrices, i.e. $A_1, A_2, ...A_n$:

```
function OptimalMatrixChainParenthesis(chain)

    n = length(chain)

    for i = 1, n

        m[i,i] = 0    // Since it takes no calculations to multiply
one matrix

    for len = 2, n

        for i = 1, n - len + 1

            j = i + len -1

          m[i,j] = infinity     // So that the first calculation updates

            for k = i, j-1

                q = m[i, k] + m[k+1, j] + p_{i-1} * p_k * p_j

                if q < m[i, j]     // The new order of parentheses is
better than what we had

                    m[i, j] = q    // Update

                    s[i, j] = k    // Record which k to split on, i.e.
where to place the parenthesis
```

So far, we have calculated values for all possible $m[i, j]$, the minimum number of calculations to multiply a chain from matrix i to matrix j, and we have recorded the corresponding "split point" $s[i,j]$. For example, if we are multiplying chain $A_1 \times A_2 \times A_3 \times A_4$, and it turns out that $m[1, 3] = 100$ and $s[1, 3] = 2$, that means that the optimal placement of parenthesis for matrices 1 to 3 is $(A_1 \times A_2) \times A_3$ and to multiply those matrices will require 100 scalar calculation.

This algorithm will produce "tables" $m[,]$ and $s[,]$ that will have entries for all possible values of i and j. The final solution for the entire chain is m[1, n], with corresponding split at s[1, n]. Unraveling the solution will be recursive, starting from the top and continuing until we reach the base case, i.e. multiplication of single matrices.

Therefore, the next step is to actually split the chain, i.e. to place the parenthesis where they (optimally) belong. For this purpose we could use the following algorithm:

```
function PrintOptimalParenthesis(s, i, j)

    if i = j

       print "A"i

    else

       print "(" PrintOptimalParenthesis(s, i, s[i, j]) PrintOptimal-
Parenthesis(s, s[i, j] + 1, j) ")"
```

Of course, this algorithm is not useful for actual multiplication. This algorithm is just a user-friendly way to see what the result looks like.

To actually multiply the matrices using the proper splits, we need the following algorithm:

```
  function MatrixChainMultiply(chain from 1 to n)        // returns the
final matrix, i.e. A1×A2×... ×An

      OptimalMatrixChainParenthesis(chain from 1 to n)    // this will
produce s[ . ] and m[ . ] "tables"

      OptimalMatrixMultiplication(s, chain from 1 to n)   // actually
multiply


  function OptimalMatrixMultiplication(s, i, j)    // returns the re-
sult of multiplying a chain of matrices from Ai to Aj in optimal way

      if i < j

         // keep on splitting the chain and multiplying the matrices
in left and right sides

         LeftSide = OptimalMatrixMultiplication(s, i, s[i, j])

         RightSide = OptimalMatrixMultiplication(s, s[i, j] + 1, j)
```

```
      return MatrixMultiply(LeftSide, RightSide)
   else if i = j
      return Ai    // matrix at position i
   else
      print "error, i <= j must hold"

function MatrixMultiply(A, B)    // function that multiplies two
matrices
   if columns(A) = rows(B)
      for i = 1, rows(A)
         for j = 1, columns(B)
            C[i, j] = 0
            for k = 1, columns(A)
                C[i, j] = C[i, j] + A[i, k]*B[k, j]
            return C
   else
       print "error, incompatible dimensions."
```

Algorithms that use Dynamic Programming

- Recurrent solutions to lattice models for protein-DNA binding.
- Backward induction as a solution method for finite-horizon discrete-time dynamic optimization problems.
- Method of undetermined coefficients can be used to solve the Bellman equation in infinite-horizon, discrete-time, discounted, time-invariant dynamic optimization problems.
- Many string algorithms including longest common subsequence, longest increasing subsequence, longest common substring, Levenshtein distance (edit distance).
- Many algorithmic problems on graphs can be solved efficiently for graphs of bounded treewidth or bounded clique-width by using dynamic programming on a tree decomposition of the graph.

- The Cocke–Younger–Kasami (CYK) algorithm which determines whether and how a given string can be generated by a given context-free grammar.
- Knuth's word wrapping algorithm that minimizes raggedness when word wrapping text.
- The use of transposition tables and refutation tables in computer chess.
- The Viterbi algorithm (used for hidden Markov models, and particularly in part of speech tagging).
- The Earley algorithm (a type of chart parser).
- The Needleman–Wunsch algorithm and other algorithms used in bioinformatics, including sequence alignment, structural alignment, RNA structure prediction.
- Floyd's all-pairs shortest path algorithm.
- Optimizing the order for chain matrix multiplication.
- Pseudo-polynomial time algorithms for the subset sum, knapsack and partition problems.
- The dynamic time warping algorithm for computing the global distance between two time series.
- The Selinger (a.k.a. System R) algorithm for relational database query optimization.
- De Boor algorithm for evaluating B-spline curves.
- Duckworth–Lewis method for resolving the problem when games of cricket are interrupted.
- The value iteration method for solving Markov decision processes.
- Some graphic image edge following selection methods such as the "magnet" selection tool in Photoshop.
- Some methods for solving interval scheduling problems.
- Some methods for solving the travelling salesman problem, either exactly (in exponential time) or approximately (e.g. via the bitonic tour).
- Recursive least squares method.
- Beat tracking in music information retrieval.

- Adaptive-critic training strategy for artificial neural networks.
- Stereo algorithms for solving the correspondence problem used in stereo vision.
- Seam carving (content-aware image resizing).
- The Bellman–Ford algorithm for finding the shortest distance in a graph.
- Some approximate solution methods for the linear search problem.
- Kadane's algorithm for the maximum subarray problem.

GLOBAL NONCONVEX OPTIMIZATION

The objective of global optimization is to find the globally best solution of (possibly nonlinear) models, in the (possible or known) presence of multiple local optima. Formally, global optimization seeks global solution(s) of a constrained optimization model. Nonlinear models are ubiquitous in many applications, e.g., in advanced engineering design, biotechnology, data analysis, environmental management, financial planning, process control, risk management, scientific modeling, and others. Their solution often requires a global search approach.

A few application examples include acoustics equipment design, cancer therapy planning, chemical process modeling, data analysis, classification and visualization, economic and financial forecasting, environmental risk assessment and management, industrial product design, laser equipment design, model fitting to data (calibration), optimization in numerical mathematics, optimal operation of "closed" (confidential) engineering or other systems, packing and other object arrangement problems, portfolio management, potential energy models in computational physics and chemistry, process control, robot design and manipulations, systems of nonlinear equations and inequalities, and waste water treatment systems management.

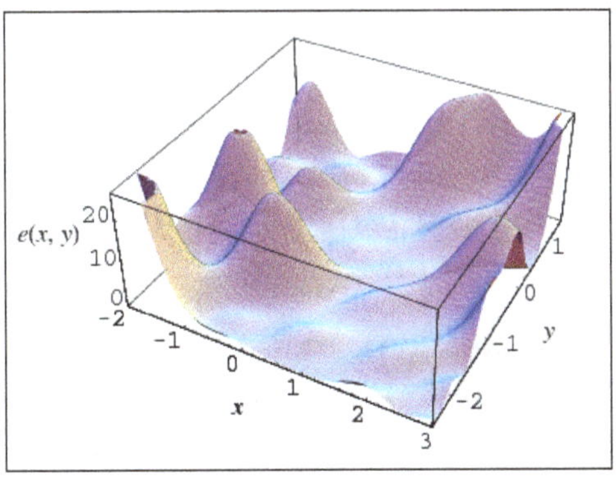

To formulate the problem of global optimization, assume that the objective function f

and the constraints g are continuous functions, the component-wise bounds x_l and x_u related to the decision variable vector x are finite, and the feasible set D is nonempty. These assumptions guarantee that the global optimization model is well-posed since, by the extreme value theorem, the solution set of the global optimization model is non-empty.

As an example, consider the l_2-norm error function related to solving the pair of equations.

$$x - \sin(2x + 3y) - \cos(3x - 5y) = 0$$
$$y - \sin(x - 2y) + \cos(x + 3y) = 0.$$

Since we wish to find the globally smallest error and not just a "locally smallest" one, we need to search globally in the two-dimensional box region. There are great many nonlinear optimization problems that possess a similar multi-modal structure.

If we use traditional local scope search methods to solve this problem, then depending on the starting point of the search, we will often find locally optimal solutions of varying quality (cf. the "valleys" in the figure above that could easily trap local search methods). In order to find the globally optimal solution, a global scope search effort is needed.

Several of the most important global optimization strategies are listed below, together with short additional remarks and references. Most global optimization software implementations are based upon one of these approaches, possibly combining ideas from several strategies.

Exact Methods

1. Naive approaches: These include the most well known passive (simultaneous) or direct (not fully adaptive) sequential global optimization strategies such as uniform grid, space covering, and pure random searches. Note that these methods are convergent under mild assumptions but, as a rule, are not practical in higher-dimensional problems.

2. Complete (enumerative) search strategies: These are based upon an exhaustive (and typically streamlined) enumeration of all possible solutions. These are applicable to combinatorial problems, as well as to certain "well-structured" continuous global optimization problems such as concave programming.

3. Homotopy (parameter continuation), trajectory methods, and related approaches: These methods have the "ambitious" objective of visiting all stationary points of the objective function: this, in turn, leads to the list of all (global as well as local) optima. This general approach includes differential equation model based, path-following search strategies, as well as fixed-point methods and pivoting algorithms.

4. Successive approximation (relaxation) methods: In this approach, the initial optimization problem is replaced by a sequence of relaxed sub-problems that are easier to solve. Successive refinement of the sub-problems to approximate the original problem is done e.g., by cutting planes and by more general cuts, diverse minorant function constructions, nested optimization and decomposition strategies and so on. These approaches are applicable to diverse structured global optimization problems such as concave minimization and differential convex problems.

5. Branch and bound algorithms: A variety of adaptive partition strategies have been proposed to solve global optimization models. These are based upon partition, sampling, and subsequent lower and upper bounding procedures: these operations are applied iteratively to the collection of active ("candidate") subsets within the feasible set *D*. Their exhaustive search feature is guaranteed in similar spirit to the analogous integer linear programming methodology. Branch and bound subsumes many specific approaches, and allows for a variety of implementations. Branch and bound methods typically rely on some a priori structural knowledge about the problem. This information may relate, for instance to how rapidly each function can vary (e.g., the knowledge of a suitable "overall" Lipschitz constant, for each function f and g); or to the availability of an analytic formulation and guaranteed smoothness of all functions (for instance, in interval arithmetic-based methods). The general branch and bound methodology is applicable to broad classes of global optimization problems, e.g., in combinatorial optimization, concave minimization, reverse convex programs, DC programming, and Lipschitz optimization.

6. Bayesian search (partition) algorithms: These methods are based upon postulated statistical information, to enable a prior stochastic description of the function-class modeled. During optimization, the problem-instance characteristics are adaptively estimated and updated. Note that, as a rule, only the corresponding one-dimensional model development is exact; furthermore, that in most practical cases "myopic" approximate decisions govern the search procedure. This general approach is applicable also to (merely) continuous global optimization problems. Theoretically, convergence to the optimal solution set is guaranteed only by generating an everywhere dense set of search points. One of the obvious challenges of using statistical methods is the choice and verification of an "appropriate" statistical model, for the class of problems to which they are applied. Additionally, it seems to be difficult to implement rigorous, computationally efficient versions of these algorithms for higher dimensional optimization problems. Note, however, that if one "skips" the underlying Bayesian paradigm, then these methods can also be pragmatically viewed as adaptive partition algorithms, and, as such, they can be directly extended to higher dimensions.

7. Adaptive stochastic search algorithms: This is another broad class of methods, based upon "exhaustive" random sampling in the feasible set. In its basic form, it includes various random search strategies that are convergent, with probability one. Search strategy adjustments, clustering and deterministic solution refinement options,

statistical stopping rules, etc. can also be added as enhancements. The methodology is applicable to both discrete and continuous global optimization problems under very mild conditions.

Heuristic Strategies

1. "Globalized" extensions of local search methods: These are partially heuristic algorithms that are nonetheless often successful in practice. The essential idea is to apply a preliminary grid search or random search based global phase, followed by applying a local (convex programming) method. For instance, random multistart performs a local search from several points selected randomly from the search domain *D*. Note that even such sampling is not trivial, when *D* has a complicated shape, as being defined, e.g., by (Lipschitz-)continuous nonlinear functions. Frequently, sophisticated algorithm enhancements are added to this basic strategy. For instance, the clustering of sample points attempts to select only a single point from each sampled "basin" of *f* from which then a local search method is initiated.

2. Evolution strategies: These methods heuristically "mimic" biological evolution: namely, the process of natural selection and the "survival of the fittest" principle. An adaptive search procedure based on a "population" of candidate solution points is used. Iterations involve a competitive selection that drops the poorer solutions. The remaining pool of candidates with higher "fitness value" are then "recombined" with other solutions by swapping components with another; they can also be "mutated" by making some smaller-scale change to a candidate. The recombination and mutation moves are applied sequentially; their aim is to generate new solutions that are biased towards subsets of *D* in which good, although not necessarily globally optimized, solutions have already been found. Numerous variants of this general strategy based on diverse evolution "game rules" can be constructed. The different types of evolutionary search methods include approaches that are aimed at continuous global optimization problems, and also others that are targeted towards solving combinatorial problems. The latter group is often called genetic algorithms.

3. Simulated annealing: These techniques are based upon the physical analogy of cooling crystal structures that spontaneously attempt to arrive at some stable (globally or locally minimal potential energy) equilibrium. This general principle is applicable to both discrete and continuous global optimization problems under mild structural requirements.

4. Tabu search: In this general category of meta-heuristics, the essential idea is to "forbid" search moves to points already visited in the (usually discrete) search space, at least for the upcoming few steps. That is, one can temporarily accept new inferior solutions, in order to avoid paths already investigated. This approach can lead to exploring new regions of D, with the goal of finding a solution by "globalized" search. Tabu search has traditionally been applied to combinatorial optimization (e.g., scheduling, routing, traveling salesman) problems. The technique can be made, at least in principle, direct-

ly applicable to continuous global optimization problems by a discrete approximation (encoding) of the problem, but other extensions are also possible.

5. Approximate convex global underestimation: This heuristically attractive strategy attempts to estimate the (postulated) large scale, "overall" convexity characteristics of the objective function f based on directed sampling in D. Applicable to smooth problems.

6. Continuation methods: These first transform the potential function into a more smooth ("simpler") function which has fewer local minimizers, and then use a local minimization procedure to trace the minimizers back to the original function. Again, this methodology is applicable to smooth problems.

7. Sequential improvement of local optima: These usually operate on adaptively constructed auxiliary functions, to assist the search for gradually better optima. The general heuristic principle includes tunneling, deflation, and filled function approaches.

References

- Optimization, science: britannica.com, Retrieved 27 April, 2019
- Battiti, Roberto; Mauro Brunato; Franco Mascia (2008). Reactive Search and Intelligent Optimization. Springer Verlag. ISBN 978-0-387-09623-0. Archived from the original on 2012-03-16
- Michael J. Todd (February 2002). "The many facets of linear programming". Mathematical Programming. 91 (3): 417–436. doi:10.1007/s101070100261. (Invited survey, from the International Symposium on Mathematical Programming.)
- Luenberger, David G.; Ye, Yinyu (2008). Linear and nonlinear programming. International Series in Operations Research & Management Science. 116 (Third ed.). New York: Springer. pp. xiv+546. ISBN 978-0-387-74502-2. MR 2423726
- GlobalOptimization: mathworld.wolfram.com, Retrieved 03 May, 2019
- Sniedovich, M. (2010), Dynamic Programming: Foundations and Principles, Taylor & Francis, ISBN 978-0-8247-4099-3

PERMISSIONS

All chapters in this book are published with permission under the Creative Commons Attribution Share Alike License or equivalent. Every chapter published in this book has been scrutinized by our experts. Their significance has been extensively debated. The topics covered herein carry significant information for a comprehensive understanding. They may even be implemented as practical applications or may be referred to as a beginning point for further studies.

We would like to thank the editorial team for lending their expertise to make the book truly unique. They have played a crucial role in the development of this book. Without their invaluable contributions this book wouldn't have been possible. They have made vital efforts to compile up to date information on the varied aspects of this subject to make this book a valuable addition to the collection of many professionals and students.

This book was conceptualized with the vision of imparting up-to-date and integrated information in this field. To ensure the same, a matchless editorial board was set up. Every individual on the board went through rigorous rounds of assessment to prove their worth. After which they invested a large part of their time researching and compiling the most relevant data for our readers.

The editorial board has been involved in producing this book since its inception. They have spent rigorous hours researching and exploring the diverse topics which have resulted in the successful publishing of this book. They have passed on their knowledge of decades through this book. To expedite this challenging task, the publisher supported the team at every step. A small team of assistant editors was also appointed to further simplify the editing procedure and attain best results for the readers.

Apart from the editorial board, the designing team has also invested a significant amount of their time in understanding the subject and creating the most relevant covers. They scrutinized every image to scout for the most suitable representation of the subject and create an appropriate cover for the book.

The publishing team has been an ardent support to the editorial, designing and production team. Their endless efforts to recruit the best for this project, has resulted in the accomplishment of this book. They are a veteran in the field of academics and their pool of knowledge is as vast as their experience in printing. Their expertise and guidance has proved useful at every step. Their uncompromising quality standards have made this book an exceptional effort. Their encouragement from time to time has been an inspiration for everyone.

The publisher and the editorial board hope that this book will prove to be a valuable piece of knowledge for students, practitioners and scholars across the globe.

INDEX

www.ingramcontent.com/pod-product-compliance
Lightning Source LLC
LaVergne TN
LVHW080642210926
848485LV00003B/307